Path Integrals
for Pedestrians

Path Integrals
for Pedestrians

Ennio Gozzi
University of Trieste & INFN, Trieste, Italy

Enrico Cattaruzza
INFN, Trieste, Italy

Carlo Pagani
INFN, Trieste, Italy & University of Mainz, Germany

World Scientific

EW JERSEY · LONDON · SINGAPORE · BEIJING · SHANGHAI · HONG KONG · TAIPEI · CHENNAI · TOKYO

Published by

World Scientific Publishing Co. Pte. Ltd.

5 Toh Tuck Link, Singapore 596224

USA office: 27 Warren Street, Suite 401-402, Hackensack, NJ 07601

UK office: 57 Shelton Street, Covent Garden, London WC2H 9HE

Library of Congress Cataloging-in-Publication Data
Gozzi, Ennio, author.
  Path integrals for pedestrians / Ennio Gozzi (University of Trieste & INFN, Italy),
Enrico Cattaruzza (INFN, Italy), Carlo Pagani (INFN, Italy & University of Mainz, Germany).
     pages cm
  Includes bibliographical references and index.
  ISBN 978-9814603928 (hardcover : alk. paper) -- ISBN 9814603929 (hardcover : alk. paper) --
  ISBN 978-9814603935 (softcover : alk. paper) -- ISBN 9814603937 (softcover : alk. paper)
  1. Path integrals. 2. Quantum theory. 3. Mechanics. I. Cattaruzza, Enrico, author.
II. Pagani, C. D., author. III. Title.
  QC174.17.P27G69 2016
  530.1201'514--dc23
                              2015029800

British Library Cataloguing-in-Publication Data
A catalogue record for this book is available from the British Library.

Cover image credit: Afro Somenzari

In-house Editor: Ng Kah Fee

Typeset by Stallion Press
Email: enquiries@stallionpress.com

Printed in Singapore

To Vitto, Patrizia and Sara

Preface

This book is intended to be a quick introduction to path integrals for people who have never before encountered them but have an undergraduate knowledge of quantum (QM) and classical mechanics (CM). We should warn the readers that this can be a *dangerous book* in the sense that, after having finished reading it, the reader may have the feeling of being able to handle path integrals and to do the relative calculations. That feeling should be put under serious control because path integrals, even if pictorially beautiful, are tricky tools that can easily lead the not-so-experienced physicist into mistakes. So we advise the reader, who really wants to embark in serious calculations, to study, after this quick introduction, the more complete books present on the market [Feynman and Hibbs (1965); Schulman (1981); Kleinert (1990); Khandekar *et al.* (1993); Swanson (1992); Wiegel (1986); Albeverio and Hoegh-Krohn (1976); Sakita (1985); Rivers (1987); Das (1993); Zinn-Justin (2005)].

The first two chapers of this book can be used, as it has been done by one of us (E. G.), for the last part of an undergraduate course on QM and can be delivered in 6 hours.

Most of the material covered in the first three chapters of the book is not original at all and it borrows heavily from [Feynman and Hibbs (1965); Schulman (1981); Khandekar *et al.* (1993); Kleinert (1990)]. The only original contents of the book, which we have not found in other books on the market, are the following two:

(1) the path integral expression for the evolution of the Wigner functions which are the quantum phase-space distributions closest to the classical ones [Marinov (1991); Gozzi and Reuter (1995)];

(2) the path-integral expression for the *operatorial* formulation of CM which was proposed in the 30's by Koopman and von Neumann

[Koopman (1931); von Neumann (1932a,b)]. It is well known that any theory which has an operatorial formulation can be put into a path integral form and we work out the case of CM [Gozzi *et al.* (1989)]. So, strangely enough, we show that CM can have a path integral formulation. This is partly the reason for the word *"pedestrians"* in the title: pedestrians are classical objects!!

The people whom we should thank are too many to be listed here and they know their names. Without their help we would have never been able to write this book.

E.G. and E.C. acknowledge financial support from grants (Prin) of the Ministry of Education and Research of Italy, from grants (Fra) of the University of Trieste, from grants of the INFN section of Trieste (Gruppo IV, IS-Genova-21 and IS-Bell), from grants of the "Consorzio per lo sviluppo della fisica" of Trieste and lately from a national INFN grant (IS-geosymqft). E.G. and E.C. warmly thank the geosymqft group of the INFN section of Naples for hosting them in their group. C.P. acknowleges the support of a grant from the "Blanceflor-Boncompagni (nee Bildt) foundation" and the hospitality of the University of Mainz.

E. Gozzi, E. Cattaruzza and C. Pagani

Contents

Preface   vii

1. The Basic Ideas   1

    1.1   Quantum mechanics and summing up amplitudes 1
    1.2   Double slit experiment 3
    1.3   Infinite slits experiment and paths correspondence 5

2. The Path Integral for Quantum Mechanics   7

    2.1   Time slicing: From infinitesimal to finite time intervals . . 7
    2.2   Re-derivation of the Feynman path integrals via the Trotter formula 8
    2.3   Continuous paths but nowhere differentiable 11
    2.4   Commutation relations 12
    2.5   Free particle 12
    2.6   Quadratic potentials and harmonic oscillator 14
    2.7   Perturbation theory via path integrals 20

3. Introduction to the Semiclassical Approximation   25

    3.1   Ordinary WKB method 26
           3.1.1   Preliminary section 26
           3.1.2   Hamilton-Jacobi equation 28
           3.1.3   WKB solutions 30
           3.1.4   Connection formulas 33
    3.2   WKB in the path integral language 34
           3.2.1   Stationary phase method 34
           3.2.2   Jacobi equation and Van Vleck determinant . . . 36

3.3   The semiclassical propagator   40
       3.3.1   Steady phase approximation method for the path
               integral   41
       3.3.2   Approximated path integral evaluation   42
       3.3.3   Functional determinants   45
       3.3.4   Final expression   46

4. Wigner Functions and its associated Path Integral   49

    4.1   Marinov's path integral for Wigner functions   53
    4.2   Semiclassical expansion in the Marinov's path integral . .   57

5. Classical Mechanics and its associated Path Integral   65

    5.1   The work of Koopman-von Neumann (KvN) on the
           operatorial version of classical mechanics   65
    5.2   Path Integrals for classical mechanics (CPI) from the KvN
           formalism   67
    5.3   Cartan calculus via the CPI   72
    5.4   Geometric quantization   76
           5.4.1   Dequantization in the q and p-polarizations and
                   supertime   81
           5.4.2   Generating functionals and Dyson-Schwinger equa-
                   tions   90
           5.4.3   Warnings on the dequantization rules   94
    5.5   Superposition in classical mechanics   99

Appendix A   Asynchronous variation of the action   105

Appendix B   The equation for the function $f(t_2, t_1)$ intro-
              duced in Section 2.6   111

Appendix C   Variational calculus in the discrete formalism   113

Appendix D   Brief review of Grassmann variables   117

Appendix E   Dimensional analysis of θ and $\bar{\theta}$ 121

Appendix F   Schrödinger and Heisenberg picture in θ and $\bar{\theta}$ 125

Appendix G   Classical path integral in the momentum rep-
    resentation   129

Appendix H   Classical path integral via the Trotter formula   133

Appendix I   Ordering problems in the classical path integral   137

Bibliography   141

Chapter 1

The Basic Ideas

1.1 Quantum mechanics and summing up amplitudes

The path integral approach to quantum mechanics was developed by R. F. Feynman in his Ph.D. thesis of 1942. It was later published (1948) in Rev. Mod. Phys. with the title *"Space-time approach to non-relativistic Quantum Mechanics"* [Feynman (1948)].

Feynman wanted a formulation of quantum mechanics in which "space-time" played a role and not just the Hilbert space, like in the traditional version of quantum mechanics. His approach is very helpful in "visualizing" many quantum mechanical phenomena and in developing various techniques, like the Feynman diagrams, the non-perturbative methods ($\hbar \to 0$, $N \to \infty$), etc. Somehow, Dirac [Dirac (1933)] had got close to the Feynman formulation of quantum mechanics in a paper in which he asked himself what is the role of the Lagrangian in quantum mechanics.

Let us first review the concept of *action* which everybody has learned in classical mechanics. Its definition is

$$S[x(t)] = \int_{(x_0, t_0)}^{(x_1, t_1)} \mathcal{L}(x(t), \dot{x}(t)) \, \mathrm{d}t \,, \tag{1.1}$$

where $x(t)$ is *any* trajectory between (x_0, t_0) and (x_1, t_1), *not* necessarily the classical one, and $\mathcal{L}$ is the Lagrangian of the system. The action $S[x(t)]$ is what in mathematical terms is known as a *functional*. Remember that a functional is a map between a space of functions $x(t)$ and a set of numbers (the real or complex numbers). From Eqn.(1.1) one sees that $S[x(t)]$ is a functional because, once we insert the function $x(t)$ on the right-hand side of Eqn.(1.1) (and perform the integration), we get a real number which is the value of the action on that trajectory. If we change the trajectory, we get a different number.

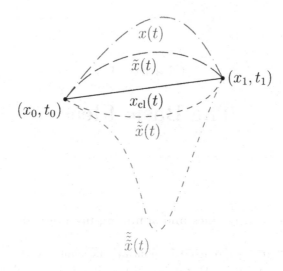

Fig. 1.1

A functional is indicated with square brackets, $S[x(t)]$, differently from a *function* whose argument is indicated with round brackets: $f(x)$. A function $f(x)$ is a map between the set of numbers (real, complex, etc.) and another set of numbers (real, complex, etc.). So, if we restrict to the real numbers, we can say that:

$$\text{\textit{Function}: } \mathbb{R} \to \mathbb{R} \,,$$
$$\text{\textit{Functional}: [functions]} \to \mathbb{R} \,.$$

Given these definitions, let us now see what the path integral formulation of quantum mechanics given by Feynman is.

We know that in quantum mechanics a central element is the *transition kernel* to go from (x_0, t_0) to (x_1, t_1) which is defined as

$$K(x_1, t_1 | x_0, t_0) \equiv \langle x_1 | e^{-i \frac{\hat{H}(t_1 - t_0)}{\hbar}} | x_0 \rangle \,. \tag{1.2}$$

What Feynman proved is the following formula:

$$K(x_1, t_1 | x_0, t_0) = \int_{(x_0, t_0)}^{(x_1, t_1)} \mathcal{D}\,[x(t)]\, e^{\frac{i}{\hbar} S[x(t)]} \,, \tag{1.3}$$

where, on the right-hand side of Eqn.(1.3), the symbol $\int_{(x_0, t_0)}^{(x_1, t_1)} \mathcal{D}\,[x(t)]$ indicates a *functional integration* which "roughly" consists of the sum over all trajectories between (x_0, t_0) and (x_1, t_1).

So, in Eqn.(1.3) we insert a trajectory in $e^{\frac{i}{\hbar}S[x(t)]}$, calculate this quantity and "sum" it to the same expression with a different trajectory and so on for all trajectories between (x_0, t_0) and (x_1, t_1). This is the reason why this method is called *path integral*. Note that all trajectories enter Eqn.(1.3) and not just the classical one.

1.2 Double slit experiment

We shall give a rigorous derivation of Eqn.(1.3) but for the moment let us try to grasp a "more physical" reason of why *trajectories* enter the expression of the quantum transition kernel. This part is taken from the book [Feynman and Hibbs (1965)].

Let us recall the double slit experiment, see Fig. 1.2.

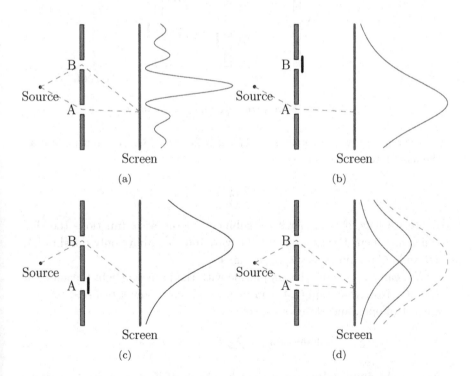

Fig. 1.2   (a) The probability P_{AB} with both slits open. (b) The probability P_A obtained with only the slit A open. (c) The probability P_B obtained keeping only the slit B open. (d) Note that $P_{AB} \neq P_A + P_B$.

In Fig. 1.2(a) both slits A and B are open while in the other two figures, 1.2(b) and 1.2(c) only one is open. It is well known that the probabilities P_{AB}, P_A, P_B satisfy the inequality

$$P_{AB} \neq P_A + P_B \,,$$

while for the probability amplitudes ψ_{AB}, ψ_A, ψ_B we have

$$\psi_{AB} = \psi_A + \psi_B \,. \tag{1.4}$$

Let us now put more screens with different openings, like in Fig. 1.3.

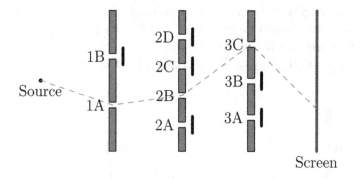

Fig. 1.3 More screens with different openings.

Let us suppose we close the slits $1B$, $2A$, $2C$, $2D$, $3A$, $3B$ and let us call the associated wave function as

$$\psi \begin{bmatrix} 1A \\ 2B \\ 3C \end{bmatrix}$$

where we have indicated with a subindex in the wave functions the slits which are open. For example for the wave function above only the slits $1A$, $2B$ and $3C$ are open as shown in Fig. 1.3.

We can "associate" this amplitude with the path that join the slits $1A$, $2B$, $3C$. Note that Eqn.(1.4) can be generalized to any set of screens with any set of open and closed slits, so:

$$\psi_{\text{(all slits open)}} = \sum_i \psi^i_{\text{(some slits closed)}} \,. \tag{1.5}$$

In turn the amplitudes on the right-hand side of Eqn.(1.5) can be written as the sum of all the amplitudes which have some of the remaining slits closed. The basic blocks of all these amplitudes will be those with only one slit open per screen and to these we can associate a *path* running through the open slits. So Eqn.(1.5) can be formally written as

$$\psi = \sum_{(\text{paths})} \psi_{(\text{path})} \, , \tag{1.6}$$

where we have substituted the labels on the wave functions with the paths which join the open slits on the various screens.

1.3 Infinite slits experiment and paths correspondence

Using this scheme Feynman had the following idea: the open space between a source and a screen can be thought of as if it were filled with an infinite set of screens each with an infinite set of slits. So now if we want the transition amplitudes from x_0 to x_1, *i.e.*, ψ_{x_0, x_1} we could write it as

$$\psi_{x_0, x_1} = \sum_{(\text{paths})} \psi_{(\text{path})} \, , \tag{1.7}$$

where the "(path)" are the labels of the amplitude associated to a configuration of screens with only one slit open through which the path passes. It is clear that the paths will be all possible paths between x_0 and x_1 because the screens have infinite slits. Let us stress that the paths in Eqn.(1.6) and (1.7) are nothing more than a *"symbol"* to indicate a set of open slits.

This somewhat gives a physical intuition of why paths—even if they are only symbols or labels—enter the transition amplitudes. Of course, one cannot say that the particles follow one path or the other because, to check that, one should do a set of measurements along the whole path while in the transition $\langle x_0, t_0 | x_1, t_1 \rangle$ the only measurements are made at x_0 and x_1. What we can say from Eqn.(1.3) is that, if we do measurements only at x_0 and x_1 then the transition amplitude is the *sum* of transition amplitudes each one associated to a different path between x_0 and x_1. If instead we do a set of infinte measurements on all points of a path to find out if the particle follows that particular path, then the interference effect between the various amplitudes entering the path integral will be destroyed.

Chapter 2

The Path Integral for Quantum Mechanics

2.1 Time slicing: From infinitesimal to finite time intervals

Let us now give a rigorous derivation of Eqn.(1.3). For small time intervals it was first done by Dirac in 1933 [Dirac (1933)], while for arbitrary time interval it was derived by Feynman in 1942 in his Ph.D. thesis which, together with the paper of Dirac, is reproduced in [Feynman (2005)].

Before proceeding to the derivation, let us review some formula regarding the exponential of operators. If $\hat{A}$ and $\hat{B}$ are two operators, then $e^{\hat{A}} e^{\hat{B}}$ is *not* equal to $e^{\hat{A}+\hat{B}}$ in general, as opposed to the case of numbers. The general formula was derived by Baker and Hausdorff and is

$$e^{\hat{A}} e^{\hat{B}} = e^{H(\hat{A},\hat{B})} , \qquad (2.1)$$

where

$$H(\hat{A}, \hat{B}) \equiv \hat{A} + \hat{B} + \frac{1}{2}[\hat{A}, \hat{B}] + \frac{1}{12}\left[\hat{A}, [\hat{A}, \hat{B}]\right] + \frac{1}{12}\left[\hat{B}, [\hat{B}, \hat{A}]\right] + \ldots \quad (2.2)$$

If $\hat{A}$ and $\hat{B}$ commute then $H(\hat{A}, \hat{B}) = \hat{A} + \hat{B}$ like in the case of c-numbers.

Let us now go back to physics and calculate the *transition kernel* which is defined as

$$K(x, t|x_0, 0) \equiv \langle x| e^{-\frac{i}{\hbar}t\hat{H}} |x_0\rangle , \qquad (2.3)$$

where $\hat{H}$ is the Hamiltonian of the system. If we divide the interval of time t into N sub-intervals we can write, using the Baker-Hausdorff formula, the following equality

$$\exp\left[-\frac{i}{\hbar}t\hat{H}\right] = \left\{\exp\left[-\frac{it}{\hbar N}\hat{H}\right]\right\}^N . \qquad (2.4)$$

This is so because the operators $t\hat{H}/N$ commute among themselves in the Baker-Hausdorff formula.

2.2 Re-derivation of the Feynman path integrals via the Trotter formula

Here we will follow the derivation contained in [Khandekar *et al.* (1993)]. Let us now note that the Hamiltonian $\hat{H}$ is the sum of two parts $\hat{H} = \hat{A} + \hat{B}$, which do not commute because $\hat{A} = \hat{p}^2/2m$ and $\hat{B} = \hat{V}(\hat{x})$. So, using again the Baker-Hausdorff formula, we can write

$$\exp\left[-\frac{i}{\hbar}\frac{t}{N}\hat{H}\right] = \exp\left[-\frac{i}{\hbar}\frac{t}{N}\left(\hat{A}+\hat{B}\right)\right]$$

$$= \exp\left[-\frac{i}{\hbar}\frac{t}{N}\hat{B}\right]\exp\left[-\frac{i}{\hbar}\frac{t}{N}\hat{A}\right] + \mathcal{O}\left(\left(\frac{t}{N}\right)^2\right),\quad (2.5)$$

where the terms $\mathcal{O}((t/N)^2)$ are those that come from the commutators of $t\hat{A}/N$ and $t\hat{B}/N$ present in Eqn.(2.2). Of course, if we take the limit $N \to \infty$ those terms are negligible with respect to the first. So combining Eqn.(2.5) with Eqn.(2.4) we get

$$\left\langle x\left|e^{-\frac{i}{\hbar}t\hat{H}}\right|x_0\right\rangle = \lim_{N\to\infty}\left\langle x\left|\left[e^{-\frac{it}{\hbar N}\hat{B}}e^{-\frac{it}{\hbar N}\hat{A}}\right]^N\right|x_0\right\rangle.\quad (2.6)$$

Let us now write down explicitly all the terms of the operator

$$\left[\exp\left(-\frac{it\hat{B}}{\hbar N}\right)\exp\left(-\frac{it}{\hbar N}\hat{A}\right)\right]^N,$$

and so Eqn.(2.6) reads

$$\left\langle x\left|\underbrace{\left[e^{-\frac{it}{\hbar N}\hat{B}}e^{-\frac{it}{\hbar N}\hat{A}}\right]\times\left[e^{-\frac{it}{\hbar N}\hat{B}}e^{-\frac{it}{\hbar N}\hat{A}}\right]\times\cdots\times\left[e^{-\frac{it}{\hbar N}\hat{B}}e^{-\frac{it}{\hbar N}\hat{A}}\right]}_{N\text{ times}}\right|x_0\right\rangle.$$

$$(2.7)$$

Next divide the interval from x_0 to x in N intervals labelled by the points $x_1, x_2, \cdots, x_{N-1}$ corresponding to the N intervals of time. Let us now insert in Eqn.(2.7), after the first square bracket, a completeness of the form $\int dx_{N-1}|x_{N-1}\rangle\langle x_{N-1}|$, where x_{N-1} is the point before x in Fig. 2.1, and continue by inserting the completeness $\int dx_{N-2}|x_{N-2}\rangle\langle x_{N-2}|$ after the second square bracket in Eqn.(2.7) and so on. What we get is that Eqn.(2.6) can be written as

$$\langle x|e^{-\frac{i}{\hbar}t\hat{H}}|x_0\rangle = \lim_{N\to\infty}\int dx_{N-1}\cdots dx_1\left\{\left\langle x\left|e^{-\frac{it}{\hbar N}\hat{B}}e^{-\frac{it}{\hbar N}\hat{A}}\right|x_{N-1}\right\rangle\right.$$

$$\left.\times\left\langle x_{N-1}\left|e^{-\frac{it}{\hbar N}\hat{B}}e^{-\frac{it}{\hbar N}\hat{A}}\right|x_{N-2}\right\rangle\times\cdots\times\left\langle x_1\left|e^{-\frac{it}{\hbar N}\hat{B}}e^{-\frac{it}{\hbar N}\hat{A}}\right|x_0\right\rangle\right\}.$$

$$(2.8)$$

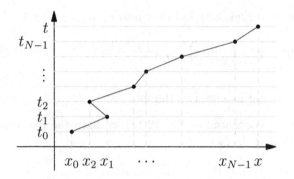

Fig. 2.1   The discretized approximation of the path integral.

All this can be put in the compact form

$$\langle x | e^{-\frac{i}{\hbar} t \hat{H}} | x_0 \rangle = \lim_{N \to \infty} \int \prod_{j=1}^{N} Q_{j,j-1} \prod_{j=1}^{N-1} \mathrm{d}x_j \,,$$

where $Q_{j,j-1}$ is

$$Q_{j,j-1} \equiv \left\langle x_j \left| e^{-\frac{it}{\hbar N} \hat{B}} e^{-\frac{it}{\hbar N} \hat{A}} \right| x_{j-1} \right\rangle .$$

In the expression above we can apply $\hat{B} = \hat{V}(\hat{x})$ to the state $\langle x_j |$ and get

$$Q_{j,j-1} = e^{-\frac{it}{\hbar N} V(x_j)} \left\langle x_j \left| e^{-\frac{it}{\hbar N} \hat{A}} \right| x_{j-1} \right\rangle . \tag{2.9}$$

The operator $\hat{A}$ is instead $\hat{p}^2/2m$ so we cannot apply it directly to the state $|x_{j-1}\rangle$. What we will do is to insert a completeness $\int |p\rangle \langle p| \, \mathrm{d}p = \mathbb{I}$ before the state $|x_{j-1}\rangle$ in Eqn.(2.9). We get in this way

$$\left\langle x_j \left| e^{-\frac{it}{\hbar N} \hat{A}} \right| x_{j-1} \right\rangle = \int \left\langle x_j \left| e^{-\frac{it}{\hbar N} \frac{\hat{p}^2}{2m}} \right| p \right\rangle \langle p | x_{j-1} \rangle \, \mathrm{d}p . \tag{2.10}$$

Remembering that

$$\langle p | x \rangle = \frac{1}{\sqrt{2\pi\hbar}} e^{-\frac{i}{\hbar} p x} \,,$$

and applying $\hat{p}^2$ on $|p\rangle$ we get from Eqn.(2.10) the following expression

$$\left\langle x_j \left| e^{-\frac{it}{\hbar N} \hat{A}} \right| x_{j-1} \right\rangle = \sqrt{\frac{1}{2\pi\hbar}} \int e^{-\frac{i}{\hbar} \left\{ \frac{t}{N} \frac{p^2}{2m} - p(x_j - x_{j-1}) \right\}} \, \mathrm{d}p$$

which, by performing the integration over p, becomes

$$\left\langle x_j \left| e^{-\frac{it}{\hbar N} \hat{A}} \right| x_{j-1} \right\rangle = \left[\frac{m}{2\pi i \hbar \frac{t}{N}} \right]^{\frac{1}{2}} e^{\frac{im}{2\hbar} \frac{N}{t} (x_j - x_{j-1})^2} . \tag{2.11}$$

Inserting this into Eqn.(2.9) and next into Eqn.(2.8) we get

$$K(x,t|x_0,0) = \lim_{N\to\infty} \left(\frac{m}{2\pi i\hbar\varepsilon}\right)^{\frac{N}{2}} \int e^{\frac{i}{\hbar}\sum_{j=1}^{N}\left\{\frac{m}{2\varepsilon}(x_j-x_{j-1})^2-\varepsilon V(x_j)\right\}} \prod_{j=1}^{N-1} dx_j \,,$$

(2.12)

where $\varepsilon = t/N$. Let us note two things:

(1) In the exponential of Eqn.(2.12) we have the discretized form of the action. In fact,

$$\int \mathcal{L} \, dt \Rightarrow \sum_j \mathcal{L}(x_j, \dot{x}_j)\varepsilon$$

$$= \sum_j \left\{\frac{m}{2}\dot{x}_j^2 - V(x_j)\right\}\varepsilon$$

$$= \sum_j \left\{m\frac{(x_j - x_{j-1})^2}{2\varepsilon^2}\varepsilon - V(x_j)\varepsilon\right\}$$

$$= \sum_j \left\{m\frac{(x_j - x_{j-1})^2}{2\varepsilon} - V(x_j)\varepsilon\right\}.$$

(2) If we look at Fig. 2.1 we see that a trajectory in the discretized form is the broken line we have drawn between x_0 and x. If we keep x_0 and x fixed and move the intermediate points (x_1, t_1), (x_2, t_2), ..., (x_{N-1}, t_{N-1}), we get all possible trajectories between x_0 and x. This is exactly what is achieved by the integrations over x_j, $j = 1, 2, \ldots, N - 1$, in Eqn.(2.12). So the sum over all trajectories indicated by the functional integration $\int \mathcal{D}\left[x(t)\right]$ in Eqn.(1.3) is actually realized by the measure of integration contained in Eqn.(2.12), *i.e.*,

$$\int \mathcal{D}\left[x(t)\right] = \lim_{N\to\infty}\left(\frac{m}{2\pi i\hbar\varepsilon}\right)^{\frac{N}{2}} \int \prod_{j=1}^{N-1} dx_j \,.$$

We reach the conclusion that Eqn.(2.12) is none other than the discretized form of the expression (1.3) and that the *functional* integral can be reduced to an infinite product of standard Lebesque integrals.

The path integral not only brings to light the role of the *non-classical* trajectories in quantum mechanics but also the role of the action in quantum mechanics. In fact the action had played a role in classical mechanics but never in quantum mechanics. This idea was the one which triggered in 1933 the work of Dirac [Dirac (1933)].

2.3 Continuous paths but nowhere differentiable

The last thing we want to bring to the attention of the reader is the indication of which are the paths which contribute most in the path integral. We shall show that these are the paths which are *continuous* but *nowhere-differentiable*.

The proof goes as follows. If we look at the kinetic piece in Eqn.(2.12) we see that in the limit of $\varepsilon \to 0$ we must have that $(x_j - x_{j-1})^2/\varepsilon$ remains finite otherwise the exponent would oscillate very fast canceling all contributions. This means

$$\frac{(x_j - x_{j-1})^2}{\varepsilon} \to \text{ finite },$$

i.e.,

$$(x_j - x_{j-1})^2 \sim \varepsilon , \qquad (\Delta x)^2 \sim \Delta t . \tag{2.13}$$

From Eqn.(2.13) we notice two things:

(1) When $\Delta t \to 0$ we have $\Delta x \to 0$, which means that the paths must be *continuous*;

(2) the velocities $\Delta x/\Delta t$ goes as $1/\Delta x$ because, using Eqn.(2.13) , we have:

$$\frac{\Delta x}{\Delta t} \sim \frac{\Delta x}{(\Delta x)^2} = \frac{1}{\Delta x} ,$$

so when $\Delta x \to 0$ the velocity has a singularity. As this happens at every point it means the path is non-differentiable.

The paths which are differentiable, *i.e.*:

$$\frac{\Delta x}{\Delta t} \to \text{ finite }$$

i.e., $\Delta x \sim \Delta t$ have a kinetic piece in the action which goes as follows:

$$\frac{(\Delta x)^2}{\Delta t} \sim \frac{(\Delta t)^2}{\Delta t} \sim \Delta t \to 0 .$$

So in the continuum $\Delta t \to 0$ these kinetic pieces go as e^0. These are constants that can be factorized out of the path integral and we can get rid of them in the normalization. So these paths do not give any phase which may interfere with the other paths and create typical quantum mechanical effects.

The *non-differentiable* paths are a typical indicator of quantum mechanical effects. In classical mechanics the paths are smooth instead. More details and examples on the path integrals can be found in the many books on the subject.

2.4   Commutation relations

The Heisenberg's uncertainty principle can be derived from the commutator

$$[\hat{x}, \hat{p}] = i\,\hbar. \tag{2.14}$$

Since in the path integrals there are only *c-numbers* and not *operators*, where does the Heisenberg's uncertainty principle emerge from? Feynman realized that there was a quantity which, if calculated at the path integral level, could give the same result as Eqn.(2.14), that quantity was the following:

$$x(t_i)\,p(t_i + \varepsilon) - p(t_i)\,x(t_i + \varepsilon). \tag{2.15}$$

If the previous expression is evaluated *inside* the path integral (we will indicate this with the symbol $\langle\dots\rangle$) detailed but simple calculations [Schulman (1981)] show that:

$$\langle x(t_i)\,p(t_i + \varepsilon) - p(t_i)\,x(t_i + \varepsilon)\rangle = i\,\hbar. \tag{2.16}$$

The result of (2.16) can be easily obtained when the *discretized form* of the path integral is used:

$$\int \mathcal{D}\,[x(t)] \exp\left\{\frac{i}{\hbar}\left[\int_{t_a}^{t_b} \mathcal{L}(x(t), \dot{x}(t)\,\mathrm{d}t\right]\right\} = \tag{2.17}$$

$$\lim_{N\to\infty}\left(\frac{m}{2\pi i\hbar\varepsilon}\right)^{\frac{N}{2}} \int \prod_{j=1}^{N-1} \mathrm{d}x_j \exp\left\{\frac{i}{\hbar}\left[\sum_j \left(m\frac{(x_j - x_{j-1})^2}{2\varepsilon} - V(x_j)\varepsilon\right)\right]\right\},$$

where $\epsilon = t/N$. The expression (2.16) is called *"time-splitting c-number commutator"* and it is the "path integral analog" of the quantum commutator in Eqn.(2.14).

2.5   Free particle

Starting from the *discretized form* of the path integral written above it is possible to get the propagator for the free particle. The result will be compared to the one obtained through the use of the *operatorial method*. The Lagrangian for a free particle is given by:

$$\mathcal{L}(\dot{x}(t)) = \frac{1}{2}m\,\dot{x}^2.$$

Inserting the expression for the Langrangian into Eqn.(2.17) we get:

$$K(x,t|x_0,0) = \lim_{N\to\infty} \left(\frac{m}{2\pi i\hbar\varepsilon}\right)^{\frac{N}{2}} \tag{2.18}$$

$$\times \int \prod_{j=1}^{N-1} \mathrm{d}x_j \exp\left\{\frac{i}{\hbar}\left[\sum_j \left(m\frac{(x_j - x_{j-1})^2}{2\varepsilon}\right)\right]\right\}.$$

So the free particle propagator is determined by a set of gaussian integrals which can be easily done, because the integral of a gaussian is a gaussian itself. Let's consider first the integral over the x_1 variable in Eqn. (2.18):

$$\left(\frac{m}{2\pi i\hbar\varepsilon}\right)^{\frac{1}{2}} \int \mathrm{d}x_1 \exp\left\{\frac{im}{2\hbar\varepsilon}\left[(x_2 - x_1)^2 + (x_1 - x_0)^2\right]\right\} =$$

$$\left(\frac{m}{2\pi i\hbar(2\varepsilon)}\right)^{\frac{1}{2}} \exp\left[\frac{im}{2\hbar(2\varepsilon)}(x_2 - x_0)^2\right].$$

Next the integration over x_2 variable has to be performed. The following contribution, taken from Eqn.(2.18), in addition to the above result has to be considered:

$$\left(\frac{m}{2\pi i\hbar\varepsilon}\right)^{\frac{1}{2}} \exp\left[\frac{im}{2\hbar\varepsilon}(x_3 - x_2)^2\right].$$

After the integration over x_2 we get:

$$\left(\frac{m}{2\pi i\hbar(3\varepsilon)}\right)^{\frac{1}{2}} \exp\left\{\frac{im}{2\hbar(3\varepsilon)}(x_3 - x_0)^2\right\}.$$

Carrying on this procedure up to the x_{N-1} integration variable, the following expression can be easily obtained:

$$\left(\frac{m}{2\pi i\hbar(N\varepsilon)}\right)^{\frac{1}{2}} \exp\left\{\frac{im}{2\hbar(N\varepsilon)}(x_N - x_0)^2\right\}.$$

Taking the limit $N \to \infty$ and $\varepsilon \to 0$ with $N\varepsilon = t$ the final result for the free-particle propagator turns out to be:

$$K(x,t|x_0,0) = \left(\frac{m}{2\pi i\hbar t}\right)^{\frac{1}{2}} \exp\left\{\frac{im}{2\hbar t}(x - x_0)^2\right\}. \tag{2.19}$$

The same result can be obtained using the *operatorial method*:

$$\langle x,t|x_0,0\rangle = \left\langle x \left| e^{-\frac{i}{\hbar}\hat{H}t} \right| x_0 \right\rangle = \left\langle x \left| e^{-\frac{it}{\hbar}\frac{\hat{p}^2}{2m}} \right| x_0 \right\rangle.$$

Inserting into the previous relation two different completeness relations $\int \mathrm{d}p\,|p\rangle\langle p| = \mathbb{I}$, we get

$$
\begin{aligned}
\langle x, t | x_0, 0 \rangle &= \int \langle x | p' \rangle \left\langle p' \left| \exp\left(-\frac{i\,t}{\hbar}\frac{\hat{p}^2}{2\,m} \right) \right| p \right\rangle \langle p | x_0 \rangle \; \mathrm{d}p \, \mathrm{d}p' \\
&= \int \frac{1}{\sqrt{2\,\pi\hbar}} \exp\left(\frac{i\,p'x}{\hbar} \right) \left\langle p' \left| \exp\left(-\frac{i\,t}{\hbar}\frac{p^2}{2\,m} \right) \right| p \right\rangle \\
&\quad \times \frac{1}{\sqrt{2\,\pi\hbar}} \exp\left(-\frac{i\,p\,x_0}{\hbar} \right) \; \mathrm{d}p \, \mathrm{d}p' \\
&= \int \frac{1}{2\,\pi\hbar} \exp\left[\frac{i\,(p'\,x - p\,x_0)}{\hbar} \right] \exp\left(-\frac{i\,t}{\hbar}\frac{p^2}{2\,m} \right) \delta(p - p') \; \mathrm{d}p \, \mathrm{d}p' \\
&= \int \frac{1}{2\,\pi\hbar} \exp\left[\frac{i\,p\,(x - x_0)}{\hbar} \right] \exp\left(-\frac{i\,t}{\hbar}\frac{p^2}{2\,m} \right) \; \mathrm{d}p.
\end{aligned}
$$

As the previous integral is a *gaussian* one, the integration can be easily performed confirming the result of Eqn.(2.19) obtained via the *path integral*.

2.6 Quadratic potentials and harmonic oscillator

In this section we will follow the derivation contained in the book of Feynman and Hibbs [Feynman and Hibbs (1965)] and the one of Schulman [Schulman (1981)]. Let's start from the following quadratic Lagrangian

$$
\mathcal{L} = a(t)\,\dot{x}^2 + b(t)\,\dot{x}\,x + c(t)\,x^2 + d(t)\,\dot{x} + e(t)\,x + f(t).
$$

The probability amplitude will be

$$
\langle x_b, t_b | x_a, t_a \rangle = \int_a^b \mathcal{D}\,[x(t)] \exp\left\{ \frac{i}{\hbar} \left[\int_{t_a}^{t_b} \mathcal{L}(x(t), \dot{x}(t), t) \, \mathrm{d}t \right] \right\} . \tag{2.20}
$$

We write the paths entering the path integral as:

$$
x(t) = x_{cl}(t) + y(t) , \tag{2.21}
$$

where $x_{cl}(t)$ is the solution of the classical equation of motion between the final and initial position x_a, x_b and the times t_a, t_b. Inserting Eqn.(2.21) into Eqn. (2.20) and using the previous Lagrangian, we get the following expression:

$$
S[x(t)] = S[x_{cl}(t) + y(t)] = \int_{t_a}^{t_b} \left[a(t)\,(\dot{x}_{cl}^2 + 2\,\dot{x}_{cl}\,\dot{y} + \dot{y}^2) + \ldots \right] \mathrm{d}t .
$$

If all the terms not containing y are collected, one obtains $S[x_{cl}]$, namely the previous action but with x_{cl} inserted. Another important thing to be stressed is that there will not be terms linear in y in the action since the classical solution is such that the first variations around it are equal to zero:

$$\frac{\delta S}{\delta x}\bigg|_{x_{cl}} = 0 . \tag{2.22}$$

It is not difficult to show that if an expansion for small values of y, up to second order, is considered in the expression for the trajectory $x(t)$, we get

$$S\left[x_{cl} + y\right] = S\left[x_{cl}\right] + \frac{\delta^2 S}{\delta x^2}\bigg|_{x_{cl}} \frac{y^2}{2!} . \tag{2.23}$$

In Eqn.(2.23) there is no contribution with $\delta^3 S/\delta x^3$ because the original S was quadratic. From the previous considerations it follows that the action can be written as:

$$S[x(t)] = S[x_{cl}(t)] + \int_{t_a}^{t_b} \left[a(t)\,\dot{y}^2 + b(t)\,\dot{y}\,y + c(t)\,y^2\right] dt . \tag{2.24}$$

Let us now go back to Eqn.(2.20) and perform the change of variable from $x(t)$ to $y(t)$ contained in Eqn.(2.21). It can be shown that the measure $\mathcal{D}\left[x(t)\right]$ is invariant:

$$\mathcal{D}\left[x(t)\right] = \mathcal{D}\left[y(t)\right], \tag{2.25}$$

because x_{cl} is a fixed path and $\mathcal{D}\left[x_{cl}\right] = 0$. If Eqn.(2.25) and Eqn. (2.24) are inserted into Eqn. (2.20) we get:

$$\langle x_b, t_b | x_a, t_a \rangle = \exp\left(\frac{i}{\hbar} S_{cl}[x_b, x_a]\right) \tag{2.26}$$

$$\times \int_0^0 \mathcal{D}\left[y(t)\right] \exp\left\{\frac{i}{\hbar}\left[\int_{t_a}^{t_b}\left[a(t)\dot{y}^2 + b(t)\,\dot{y}\,y + c(t)y^2\right] dt\right]\right\}.$$

It has been possible to pull out the phase $\exp\left(\frac{i}{\hbar} S_{cl}[x_b, x_a]\right)$ from the path integral because it does not depend on y. The rest of the path integral contains the y trajectories which go from $y = 0$ at the time t_a to $y = 0$ at the time t_b. The extrema of integration of y can be derived from the analysis of Fig. 2.2. Note that the trajectories in y entering the path integral are periodic and this fact simplifies the evaluation of the path integral itself. Looking at the expression of Eqn.(2.26) one notices that the second piece does not depend on x_a, x_b anymore. The dependence on x_a and x_b is present only in the phase $\exp\left(\frac{i}{\hbar} S_{cl}[x_b, x_a]\right)$ since it is a function of x_{cl}, the classical trajectory from x_a to x_b. In the path integral piece the dependence on t_a and t_b will survive because the trajectories in y go from 0 to 0 in the time interval $t_b - t_a$. Eqn.(2.26) will then have the general form:

$$\langle x_b, t_b | x_a, t_a \rangle = \exp\left(\frac{i}{\hbar} S_{cl}[x_b, x_a]\right) F(t_a, t_b), \tag{2.27}$$

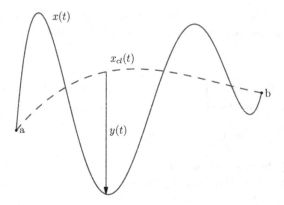

Fig. 2.2   A trajectory in $y(t)$ which indicates the deviation from the classical path.

where $F(t_a, t_b)$ is a function to be determined. Let's consider as an example the harmonic oscillator:

$$\mathcal{L} = \frac{m}{2}\dot{x}^2 - \frac{m\omega^2}{2}x^2.$$

In Eqn.(2.26) the first thing to do is to calculate the quantity $S_{cl}[x_b, x_a]$, where S_{cl} is the action on the classical path between x_a and x_b. We can easily solve the equation of motion for the harmonic oscillator with boundary conditions $(x_a, t_a), (x_b, t_b)$. Suppose we indicate the classical solution as $x_{cl}(t) = f(t_b - t_a, x_a, x_b, t)$. Then inserting this expression into S and integrating over t, the following expression can be found:

$$S_{cl} = \frac{m\omega}{2\sin(\omega T)}\left[(x_a^2 + x_b^2)\cos(\omega T) - 2\,x_a\,x_b\right],$$

where $T = t_b - t_a$. Next the quantity $F(T) = F(t_b, t_a)$ of Eqn. (2.27) has to be evaluated:

$$F(T) = \int_{0,0}^{0,T} \mathcal{D}\,[y(t)]\exp\left[\frac{i}{\hbar}\int_0^T \frac{m}{2}\left(\dot{y}^2 - \omega^2\,y^2\right)dt\right]. \qquad (2.28)$$

This integral can be calculated by *discretizing* the time interval T and considering the periodic nature of the trajectories entering the path integral. They are in fact periodic functions of period T ($y(0) = y(T) = 0$) and can therefore be expressed as:

$$y(t) = \sum_n a_n \sin\left(\frac{n\pi t}{T}\right). \qquad (2.29)$$

In this way the functional *"variability"* of $y(t)$ has been reduced to the freedom of varying the set of numbers a_n. The functional integration over $y(t)$ becomes a set of Lebesgue integration over the a_n:

$$\int_{0,0}^{0,T} \mathcal{D}\left[y(t)\right] = J \int_{-\infty}^{+\infty} \cdots \int_{-\infty}^{+\infty} \prod_{n=1}^{\infty} \mathrm{d}a_n,$$

where J is the Jacobian of the transformation from $y(t)$ to the infinite set $\{a_n\}_{n\in\mathbb{N}}$. If Eqn.(2.29) is inserted into Eqn.(2.28) the following contributions can be obtained for the kinetic part

$$\int \dot{y}^2\,\mathrm{d}t = \sum_n \sum_m \left(\frac{n\pi}{T}\right)\left(\frac{m\pi}{T}\right) a_n\, a_m \int_0^T \cos\left(\frac{n\pi t}{T}\right)\cos\left(\frac{m\pi t}{T}\right)\mathrm{d}t$$

$$= \frac{T}{2}\sum_n \left(\frac{n\pi}{T}\right)^2 a_n^2$$

and for the potential one:

$$\int y^2\,\mathrm{d}t = \sum_n \sum_m a_n\, a_m \int_0^T \sin\left(\frac{n\pi t}{T}\right)\sin\left(\frac{m\pi t}{T}\right)\mathrm{d}t$$

$$= \frac{T}{2}\sum_n a_n^2.$$

Inserting the previous expressions into Eqn.(2.28) we obtain

$$F(T) = J \int_{-\infty}^{+\infty} \cdots \int_{-\infty}^{+\infty} \prod_{n=1}^{\infty} \mathrm{d}a_n \exp\left\{\frac{imT}{4\hbar}\sum_n\left[\left(\frac{n\pi}{T}\right)^2 - \omega^2\right]a_n^2\right\}.$$

$$(2.30)$$

In the previous formula the a_n do not get coupled to each other, so it is possible to perform separately each *gaussian* integration:

$$\int_{-\infty}^{+\infty} \exp\left\{\frac{imT}{4\hbar}\left[\left(\frac{n\pi}{T}\right)^2 - \omega^2\right]a_n^2\right\}\mathrm{d}a_n \propto \left[\left(\frac{n\pi}{T}\right)^2 - \omega^2\right]^{-\frac{1}{2}}.$$

In this manner we obtain the following result:

$$F(T) = J \prod_{n=1}^{\infty}\left[\int_{-\infty}^{+\infty}\mathrm{d}a_n \exp\left\{\frac{imT}{4\hbar}\left[\left(\frac{n\pi}{T}\right)^2 - \omega^2\right]a_n^2\right\}\right]$$

$$\propto \prod_{n=1}^{\infty}\left[\left(\frac{n\pi}{T}\right)^2 - \omega^2\right]^{-\frac{1}{2}} = \prod_{n=1}^{\infty}\left[\left(\frac{n\pi}{T}\right)^2\right]^{-\frac{1}{2}}\prod_{n=1}^{\infty}\left[1 - \left(\frac{\omega T}{n\pi}\right)^2\right]^{-\frac{1}{2}}.$$

$$(2.31)$$

The second term above is equal to:

$$\prod_{n=1}^{\infty}\left[1-\left(\frac{\omega T}{n\pi}\right)\right]^{-\frac{1}{2}} = \left\{\prod_{n=1}^{\infty}\left[1-\left(\frac{\omega T}{n\pi}\right)^2\right]\right\}^{-\frac{1}{2}} = \left[\frac{\sin(\omega T)}{\omega T}\right]^{-\frac{1}{2}}.$$

The first term in Eqn. (2.31) does not depend on ω which contains all the dynamical information so it can be absorbed into a constant C entering the whole expression for $F(T)$. Therefore we can write:

$$F(T) = C\left[\frac{\sin(\omega T)}{\omega T}\right]^{-\frac{1}{2}}. \tag{2.32}$$

How can we determine the constant C? The expression of $F(T)$ is known for the free particle (see Eqn.(2.19)), which is obtained in the limit $\omega \to 0$

$$\lim_{\omega\to 0} F(T) = \left(\frac{m}{2\pi i\hbar T}\right)^{\frac{1}{2}}.$$

So this is the value of C, therefore:

$$F(T) = \left[\frac{m\omega}{2\pi i\hbar\,\sin(\omega T)}\right]^{\frac{1}{2}}. \tag{2.33}$$

The complete propagator for the harmonic oscillator is then given by:

$$\langle x_b, t_b | x_a, t_a \rangle = \exp\left(\frac{i}{\hbar}\frac{m\omega}{2\sin(\omega T)}\left[(x_a^2 + x_b^2)\cos(\omega T) - 2x_a x_b\right]\right)$$
$$\times\left[\frac{m\omega}{2\pi i\hbar\,\sin(\omega T)}\right]^{\frac{1}{2}}.$$

Let us provide a further derivation of the results just obtained. We want to evaluate the expression (2.28) starting from the discretized form of the path integral. The discretized action is given by (see Appendix C for the general case)[1]:

$$\tilde{S}_N = \frac{m}{2}\varepsilon\sum_{k=1}^{N+1}\left[\left(\frac{\eta_{\alpha_k} - \eta_{\alpha_{k-1}}}{\varepsilon}\right)^2 - \omega^2\eta_{\alpha_k}^2\right]. \tag{2.34}$$

We define the N-component vector

$$\eta \equiv \begin{pmatrix} \eta_{\alpha_1} \\ \eta_{\alpha_2} \\ \vdots \\ \eta_{\alpha_N} \end{pmatrix} \tag{2.35}$$

---

[1]We have $\eta_{\alpha_0} = \eta_{\alpha_{N+1}} = 0$.

and the $N \times N$ matrix

$$\underline{\underline{\sigma}} \equiv \frac{i}{\hbar} \frac{m}{2} \frac{1}{\varepsilon} \left[\begin{pmatrix} -2 & 1 & & & & 0 \\ 1 & -2 & 1 & & & \\ & 1 & -2 & 1 & & \\ & & \ddots & \ddots & \ddots & \\ & & & 1 & -2 & 1 \\ 0 & & & & 1 & -2 \end{pmatrix} + \varepsilon\omega^2 \begin{pmatrix} 1 & & & 0 \\ & 1 & & \\ & & 1 & \\ & & & \ddots \\ & & & & 1 \\ 0 & & & & 1 \end{pmatrix} \right]. \quad (2.36)$$

We can then rewrite the function $F(t_a, t_b)$ as follows

$$\lim_{N\to\infty} \frac{1}{(2\pi\hbar i\varepsilon/m)^{(N+1)/2}} \int d\eta \, e^{-\eta^T \underline{\underline{\sigma}} \eta}. \quad (2.37)$$

Let us observe that the matrix $\underline{\underline{\sigma}}$ is Hermitian and can be brought into diagonal form via a unitary transformation. Let $\underline{\underline{U}}$ be the eigenvalue matrix and $\zeta = \underline{\underline{U}}^{-1}\eta$ the new integration variables, we can then get:

$$\int d\eta \, e^{-\eta^T \underline{\underline{\sigma}} \eta} = \int d\zeta \, e^{-\zeta^T \underline{\underline{U}}^T \underline{\underline{\sigma}} \, \underline{\underline{U}}\zeta} = \int d\zeta \, e^{-\zeta^T \underline{\underline{\Lambda}}\zeta}$$

$$= \int_{-\infty}^{+\infty} d\zeta_{\alpha_1} \int_{-\infty}^{+\infty} d\zeta_{\alpha_2} \cdots \int_{-\infty}^{+\infty} d\zeta_{\alpha_N} \exp\left(-\sum_{k=1}^{N} \lambda_k \zeta_{\alpha_k}^2 \right)$$

$$= \prod_{k=1}^{N} \sqrt{\frac{\pi}{\lambda_k}} = \sqrt{\frac{\pi^N}{\det \underline{\underline{\sigma}}}}. \quad (2.38)$$

Hence we find

$$F(t_a, t_b) = \lim_{N\to\infty} \sqrt{\left(\frac{m}{2\pi\hbar i\varepsilon}\right)^{N+1} \frac{\pi^N}{\det \underline{\underline{\sigma}}}}. \quad (2.39)$$

Following [Schulman (1981)] we define

$$f(t_2, t_1) \equiv \lim_{N\to\infty} \varepsilon \left(\frac{2\hbar i\varepsilon}{m}\right)^N \det \underline{\underline{\sigma}} \quad (2.40)$$

and rewrite

$$F(t_a, t_b) = \sqrt{\frac{m}{2\pi\hbar i f(T,0)}}. \quad (2.41)$$

Performing the formal manipulations which can be found in Appendix B, it is possible to show that, in the limit $N \to \infty$ and $\varepsilon \to 0$, the function $f(t_2, t_1)$ satisfies the differential equations:

$$\begin{cases} m\dfrac{\partial^2 f(t_2, t_1)}{\partial t_2^2} + m\omega^2 f(t_2, t_1) = 0 \\[4mm] f(t_1, t_1) = 0 \text{ and } \dfrac{\partial f(t_2, t_1)}{\partial t_2}\bigg|_{t_2=t_1} = 1. \end{cases} \quad (2.42)$$

The solution reads

$$f(t_2, t_1) = \frac{\sin \omega (t_2 - t_1)}{\omega} = \frac{\sin \omega T}{\omega}. \qquad (2.43)$$

If we now insert the above expression into Eqn.(2.41) we recover the result (2.33).

2.7   Perturbation theory via path integrals

In this section we will follow [Feynman and Hibbs (1965)] and derive the time-dependent perturbative expansion for the time-evolution kernel via path integrals.

We have seen that the transition kernel between $x = x_a$ and $x = x_b$ from time t_a to time t_b has the following path integral expression:

$$K_V(b, a) = \int_a^b \mathcal{D}\left[x(t)\right] \exp\left\{\frac{i}{\hbar}\left[\int_{t_a}^{t_b}\left(\frac{m\dot{x}^2}{2} - V(x, t)\right) dt\right]\right\}. \qquad (2.44)$$

If we suppose that the potential is small or better yet $\int_{t_a}^{t_b} V(x, t)\, dt$ is small with respect to $\hbar$, then we can expand the integrand in Eqn.(2.44) as follows:

$$\exp\left[-\frac{i}{\hbar}\int_{t_a}^{t_b} V(x, t)\, dt\right] = 1 - \frac{i}{\hbar}\int_{t_a}^{t_b} V(x, t)\, dt$$

$$+ \frac{1}{2!}\left(\frac{i}{\hbar}\right)^2 \left[\int_{t_a}^{t_b} V(x, t)\, dt\right]^2 + \dots \quad (2.45)$$

Of course, $\int_{t_a}^{t_b} V(x, t)\, dt$ is a functional, so the statement that $\int_{t_a}^{t_b} V(x, t)\, dt$ should be "small" with respect to $\hbar$ needs to be better clarified. We will leave this clarification for more advanced courses.

Using the above expansion, we can rewrite $K_V(b, a)$ in the following way:

$$K_V(b, a) = K^{(0)}(b, a) + K^{(1)}(b, a) + K^{(2)}(b, a) + \dots, \qquad (2.46)$$

where

$$K^{(0)}(b, a) = \int_a^b \mathcal{D}\left[x(t)\right]\left[\exp\left(\frac{i}{\hbar}\int_{t_a}^{t_b}\frac{m\dot{x}^2}{2}\, dt\right)\right],$$

$$K^{(1)}(b, a) = -\frac{i}{\hbar}\int_a^b \mathcal{D}\left[x(t)\right]\left\{\left(\int_{t_a}^{t_b} V(x(s), s)\, ds\right)\exp\left[\frac{i}{\hbar}\int_{t_a}^{t_b}\frac{m\dot{x}^2}{2}\, dt\right]\right\},$$

$$K^{(2)}(b, a) = -\frac{1}{2\hbar^2}\int_a^b \mathcal{D}\left[x(t)\right]\left\{\left[\int_{t_a}^{t_b} V(x(s), s)\, ds\right]\left[\int_{t_a}^{t_b} V(x(s'), s')\, ds'\right]\right.$$

$$\left. \times \exp\left[\frac{i}{\hbar}\int_{t_a}^{t_b}\frac{m\dot{x}^2}{2}\, dt\right]\right\}.$$

Let us now proceed to evaluate the various terms.

$K^{(0)}(b,a)$ is the free particle transition kernel. $K^{(1)}(b,a)$ can be written as

$$K^{(1)}(b,a) = -\frac{i}{\hbar} \int_{t_a}^{t_b} F(s) \, \mathrm{d}s \, , \qquad (2.47)$$

where

$$F(s) = \int_a^b \mathcal{D}\left[x(t)\right] V(x(s),s) \exp\left(\frac{i}{\hbar} \int_{t_a}^{t_b} \frac{m\dot{x}^2}{2} \, \mathrm{d}t\right) \, . \qquad (2.48)$$

Basically, $F(s)$ is the path integral of the free particle but with the potential inserted at $t = s$. So the time evolution before $t = s$ is the one of the free particle, at $t = s$ it gets "perturbed" by $V(x(s),s)$, afterwards it is again the evolution of the free particle. The picture that we can associate to Eqn.(2.48) is the one in Fig. 2.3. There, we have indicated the point

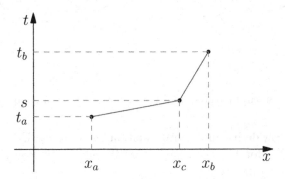

Fig. 2.3   Schematic representation of $F(s)$.

reached at time s as x_c. In Eqn.(2.48) we are effectively integrating over the x_c and as a consequence $F(s)$ can be written in the following manner:

$$F(t_c) = \int_{-\infty}^{+\infty} K^{(0)}(b,c)V(x_c,t_c)K^{(0)}(c,a) \, \mathrm{d}x_c \, .$$

Using this expression, $K^{(1)}(b,a)$ can be represented as

$$K^{(1)}(b,a) = -\frac{i}{\hbar} \int_{t_a}^{t_b} \int_{-\infty}^{+\infty} K^{(0)}(b,c)V(x_c,t_c)K^{(0)}(c,a) \, \mathrm{d}x_c \, \mathrm{d}t_c \, . \qquad (2.49)$$

The pictures that can be drawn associated to the perturbation series of Eqn.(2.46) are, in analogy to that in Fig. 2.3, the ones in Fig. 2.4. These are what we would call *proto-Feynman diagrams* in a potential theory.

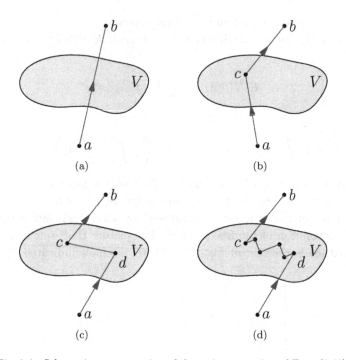

Fig. 2.4   Schematic representation of the series expansion of Eqn. (2.46).

We shall now derive an integral equation for $K_V(b,a)$. Using Eqn.(2.49) and the analog for higher orders we easily get the following expression (where $\int \mathrm{d}\tau_c = \int \mathrm{d}x_c\,\mathrm{d}t_c$):

$$K_V(b,a) = K^{(0)}(b,a) - \frac{i}{\hbar} \int K^{(0)}(b,c)V(x_c,t_c)K^{(0)}(c,a)\,\mathrm{d}\tau_c$$

$$+ \left(-\frac{i}{\hbar}\right)^2 \int\int K^{(0)}(b,c)V(x_c,t_c)K^{(0)}(c,d)V(x_d,t_d)K^{(0)}(d,a)\,\mathrm{d}\tau_c\,\mathrm{d}\tau_d$$

$$+ \cdots , \quad (2.50)$$

which can also be written as

$$K_V(b,a) = K^{(0)}(b,a) - \frac{i}{\hbar} \int K^{(0)}(b,c)V(x_c,t_c)$$

$$\times \left[K^{(0)}(c,a) - \frac{i}{\hbar} \int K^{(0)}(c,d)V(x_d,t_d)K^{(0)}(d,a)\,\mathrm{d}\tau_d + \ldots \right] \mathrm{d}\tau_c. \quad (2.51)$$

If we look at Eqn.(2.51) and in particular at the quantity inside the square brackets, we notice that Eqn.(2.51) can be rewritten as

$$K_V(b,a) = K^{(0)}(b,a) - \frac{i}{\hbar} \int K^{(0)}(b,c)V(x_c,t_c)K_V(c,a)\,\mathrm{d}\tau_c\ . \qquad (2.52)$$

This is an integral equation for K_V. It is completely equivalent to the differential Schrödinger equation: it is basically the *integral* version of it.

Chapter 3

Introduction to the Semiclassical Approximation

The path integral[1] is not only the best tool for perturbation theory[2] but also a very powerful instrument to develop the so-called *non-perturbative techniques*. One of these is the semiclassical approximation. In this chapter we will briefly review it as it is done in most textbooks of quantum mechanics and then present a version involving path integrals.

In the ordinary formulation of quantum mechanics the central problem is to solve the Schrödinger equation

$$\left[-\frac{\hbar^2}{2m} \nabla^2 + V\left(\mathbf{r}, t\right) \right] \Psi\left(\mathbf{r}, t\right) = i\hbar \frac{\partial}{\partial t} \Psi\left(\mathbf{r}, t\right) \tag{3.1}$$

with given conditions on the wave function, e.g. the wave function normalization $\Psi\left(\mathbf{r}, t\right)$ and the initial condition $\Psi\left(\mathbf{r}, 0\right) = \Psi_0$. This problem does not often find an exact solution because of the mathematical difficulties arising for several potentials. Many approximation methods have been developed to simplify the differential equation (3.1) and to look for solutions that, although approximated, are sufficient in order to extract informations regarding the physical system under study. The approximation method we will review here is the so-called *semiclassical method* or *WKB method*. The idea at the basis of this method is to consider an expansion not in the interaction coupling but in the Planck constant $\hbar$. Since $\hbar$ is a dimensional quantity, it is the limit of the dimensionless quantity $\hbar/S_{cl} \to 0$ which has to be taken into account (S_{cl} is the action calculated on a classical path). The limit $\hbar \to 0$ should reduce quantum quantities to the classical ones according to the *Correspondence Principle*. But this does not mean that

---

[1]The content of this chapter is based on an undergraduate thesis by A. Taracchini, written in the year 2006.

[2]Note that the Feynman diagrams have been first derived by Feynman from what would later be known as the path integral method.

quantum mechanics (QM) can be reduced to classical mechanics (CM). In fact the limit $\hbar \to 0$ of QM contains some *phases*, which will remain also in this limit. The expression of the physical quantities of interest will be different from the exact quantum ones and it has the advantage that they can be reconstructed starting from the classical information. Manipulating the Schrödinger equation (3.1), it is possible to find two equations that involve the phase and the modulus of the wavefunction. At the zeroth order in $\hbar^2$ the first equation for the phase turns out to be nothing else than the *classical Hamilton-Jacobi equation*, while the second is the equation for the modulus. In these two equations at the zeroth order in $\hbar^2$ the phase and the modulus are decoupled from each other. This is a typical peculiarity of classical mechanics. At the quantum level, instead, the phase and the modulus never decouple from each other and this is the central feature of quantum mechanics. The WKB method goes beyond the zeroth order in $\hbar^2$. This approximation, as we will see later on, is very good if the potential varies slowly in space. In this chapter the semiclassical method will be presented both in the standard formulation of quantum mechanics and also in the path integral one [Schulman (1981)]. In this last formulation the evaluation of the propagator will be done summing up over the classical trajectories and it will be possible, like in the ordinary WKB, to recover not only the classical equations of Hamilton-Jacobi but also, going to higher orders, other classical equations like the ones for the first variations or *Jacobi fields* around the classical trajectories. It is particularly interesting to apply the WKB method to those phenomena whose nature is exclusively quantum. A typical example is represented by the tunneling effect occurring in a double well potential. In this case there are no classical trajectories with which to build the semiclassical expressions. The path integral method is crucial in understanding which pseudo-classical elements enter the tunnel effect: these are classical solutions of the imaginary-time action called *instantons*, which were discovered for the first time in 1975 [Belavin *et al.* (1975)].

3.1 Ordinary WKB method

3.1.1 *Preliminary section*

In general the wavefunction associated to a physical system, being a complex quantity, can be written as:

$$\Psi\left(\mathbf{r},t\right) = A\left(\mathbf{r},t\right)\exp\left[\frac{i}{\hbar}S\left(\mathbf{r},t\right)\right], \tag{3.2}$$

where $A\left(\mathbf{r},t\right)$ and $S\left(\mathbf{r},t\right)$ are real functions. Let us substitute Eqn.(3.2) into the Schrödinger equation (3.1) and let us evaluate first the piece:

$$\begin{aligned}
\nabla^2\Psi\left(\mathbf{r},t\right) &= \nabla^2\left\{A\left(\mathbf{r},t\right)\exp\left[\frac{i}{\hbar}S\left(\mathbf{r},t\right)\right]\right\} \\
&= \left\{\nabla^2 A\left(\mathbf{r},t\right) + \frac{2i}{\hbar}\nabla A\left(\mathbf{r},t\right)\cdot\nabla S\left(\mathbf{r},t\right) + \frac{i}{\hbar}A\left(\mathbf{r},t\right)\nabla^2 S\left(\mathbf{r},t\right)\right. \\
&\quad\left. -\frac{1}{\hbar^2}A\left(\mathbf{r},t\right)\left[\nabla S\left(\mathbf{r},t\right)\right]^2\right\}\exp\left[\frac{i}{\hbar}S\left(\mathbf{r},t\right)\right]
\end{aligned} \tag{3.3}$$

and then the term

$$\begin{aligned}
\frac{\partial}{\partial t}\Psi\left(\mathbf{r},t\right) &= \frac{\partial}{\partial t}\left\{A\left(\mathbf{r},t\right)\exp\left[\frac{i}{\hbar}S\left(\mathbf{r},t\right)\right]\right\} \\
&= \left\{\frac{\partial}{\partial t}A\left(\mathbf{r},t\right) + \frac{i}{\hbar}A\left(\mathbf{r},t\right)\frac{\partial}{\partial t}S\left(\mathbf{r},t\right)\right\}\exp\left[\frac{i}{\hbar}S\left(\mathbf{r},t\right)\right].
\end{aligned} \tag{3.4}$$

Using Eqn.(3.3) and Eqn.(3.4), the Schrödinger equation becomes:

$$\begin{aligned}
&-\frac{\hbar^2}{2m}\nabla^2 A\left(\mathbf{r},t\right) - \frac{i\hbar}{m}\nabla A\left(\mathbf{r},t\right)\cdot\nabla S\left(\mathbf{r},t\right) - \frac{i\hbar}{2m}A\left(\mathbf{r},t\right)\nabla^2 S\left(\mathbf{r},t\right) \\
&+\frac{1}{2m}A\left(\mathbf{r},t\right)\left[\nabla S\left(\mathbf{r},t\right)\right]^2 + V\left(\mathbf{r},t\right)A\left(\mathbf{r},t\right) \\
&= i\hbar\frac{\partial}{\partial t}A\left(\mathbf{r},t\right) - A\left(\mathbf{r},t\right)\frac{\partial}{\partial t}S\left(\mathbf{r},t\right).
\end{aligned} \tag{3.5}$$

Comparing real and imaginary parts of Eqn.(3.5), the following two relations can be obtained:

$$\frac{\partial}{\partial t}S\left(\mathbf{r},t\right) + \frac{\left[\nabla S\left(\mathbf{r},t\right)\right]^2}{2m} + V\left(\mathbf{r},t\right) = \frac{\hbar^2}{2m}\frac{\nabla^2 A\left(\mathbf{r},t\right)}{A\left(\mathbf{r},t\right)} \tag{3.6a}$$

$$m\frac{\partial}{\partial t}A\left(\mathbf{r},t\right) + \nabla A\left(\mathbf{r},t\right)\cdot\nabla S\left(\mathbf{r},t\right) + \frac{1}{2}A\left(\mathbf{r},t\right)\nabla^2 S\left(\mathbf{r},t\right) = 0. \tag{3.6b}$$

The second equation is the *continuity equation* for the probability density. It is well known that the probability density is defined as follows

$$\rho \equiv \left|\Psi\left(\mathbf{r},t\right)\right|^2 = A\left(\mathbf{r},t\right)^2 \tag{3.7}$$

and the probability current as:

$$\mathbf{J} \equiv \mathrm{Re}\left[\Psi^{*}\left(\mathbf{r},t\right)\frac{\hbar}{im}\nabla\Psi\left(\mathbf{r},t\right)\right] = A\left(\mathbf{r},t\right)^{2}\frac{\nabla S\left(\mathbf{r},t\right)}{m}, \qquad (3.8)$$

then Eqn. (3.6b) becomes

$$\frac{\partial\rho}{\partial t} + \nabla \cdot \mathbf{J} = 0. \qquad (3.9)$$

Moreover let us note that Eqn.(3.6a), at the zeroth order in $\hbar^2$, becomes the *classical Hamilton-Jacobi equation*[3]

$$\frac{\partial}{\partial t}S_0\left(\mathbf{r},t\right) + \frac{\left[\nabla S_0\left(\mathbf{r},t\right)\right]^2}{2m} + V\left(\mathbf{r},t\right) = 0. \qquad (3.10)$$

Let us review now the principal features of the previous equation in the context of Classical Mechanics.

3.1.2 *Hamilton-Jacobi equation*

It is well known [Goldstein (2002)] that the fundamental equations of the Hamiltonian formulation of classical mechanics give the evolution of the *canonical variables* $x_1, ..., x_{2f}$, namely the Lagrangian coordinates $q_1, ..., q_f$ and the conjugate momenta $p_1, ..., p_f$, and have the form[4]

$$\dot{q}_k = \{q_k, H\}_{PB} \qquad (3.11a)$$

$$\dot{p}_k = \{p_k, H\}_{PB} \qquad (3.11b)$$

with $k = 1, ..., f$, for a system with f degrees of freedom and Hamiltonian H. The variables $x_1, ..., x_{2f}$ will be indicated with the different symbol $\varphi_1, ..., \varphi_{2f}$ in the next chapter. It is possible to introduce some particular transformations of the variables in the phase space

$$Q_k = Q_k\left(q_1, ..., q_f, p_1, ..., p_f, t\right) \qquad (3.12a)$$

$$P_k = P_k\left(q_1, ..., q_f, p_1, ..., p_f, t\right) \qquad (3.12b)$$

which leave invariant in form the Hamilton equations. These transformations are called *canonical* transformations. We will indicate with $X_1, ..., X_{2f}$

---

[3] We have written $S_0\left(\mathbf{r},t\right)$ and not simply $S\left(\mathbf{r},t\right)$ to underline the fact that we are working in the zeroth approximation (with respect to $\hbar^2$).

[4] The $\{,\}_{PB}$ are the Poisson brackets defined as:

$$\{F,G\}_{PB} \equiv \sum_{k=1}^{f}\left(\frac{\partial F}{\partial q_k}\frac{\partial G}{\partial p_k} - \frac{\partial F}{\partial p_k}\frac{\partial G}{\partial q_k}\right),$$

where $F = F\left(q_1, ..., q_f, p_1, ..., p_f, t\right)$ and $G = G\left(q_1, ..., q_f, p_1, ..., p_f, t\right)$ are two arbitrary functions of the phase space variables.

the new phase space variables and with K the new Hamiltonian. The fundamental condition that has to be satisfied in order to have a *canonical transformation* is the following (known as *sympleptic* condition):

$$\underline{\underline{M}} \; \underline{\underline{J}} \; \underline{\underline{M}}^T = \underline{\underline{J}} \tag{3.13}$$

where $\underline{\underline{M}}$ is the Jacobian matrix of the transformation[5]

$$\underline{\underline{M}} \equiv \frac{\partial \mathbf{X}}{\partial \mathbf{x}} = \frac{\partial (\mathbf{Q}, \mathbf{P})}{\partial (\mathbf{q}, \mathbf{p})} = \frac{\partial (Q_1, ..., Q_f, P_1, ..., P_f)}{\partial (q_1, ..., q_f, p_1, ..., p_f)} \tag{3.14}$$

and $\underline{\underline{J}}$ is called the Poisson matrix and it isdefined as:

$$\underline{\underline{J}} \equiv \begin{pmatrix} \underline{\underline{0}}_{f \times f} & \underline{\underline{1}}_{f \times f} \\ -\underline{\underline{1}}_{f \times f} & \underline{\underline{0}}_{f \times f} \end{pmatrix}_{2f \times 2f} . \tag{3.15}$$

According to Hamilton variational principle, the solution of Eqn.(3.11a) and Eqn.(3.11b) satisfies the variational principle:

$$\delta \int_{t_1}^{t_2} dt \, L \left(q_1, ..., q_f, \dot{q}_1, ..., \dot{q}_f, t \right)$$

$$= \delta \int_{t_1}^{t_2} dt \left[\sum_{k=1}^{f} \dot{q}_k p_k - H \left(q_1, ..., q_f, p_1, ..., p_f, t \right) \right] = 0. \tag{3.16}$$

An analogous expression will be valid for the new variables in phase space given by Eqn.(3.12a) and Eqn.(3.12b), that is

$$\delta \int_{t_1}^{t_2} dt \left[\sum_{k=1}^{f} \dot{Q}_k P_k - K \left(Q_1, ..., Q_f, P_1, ..., P_f, t \right) \right] = 0. \tag{3.17}$$

The two equations Eqn.(3.16) and Eqn.(3.17) describe the same physical system if and only if the symplectic condition (3.13) is satisfied and if the integrands of the two equations above differ by a total derivative of a function $F(\mathbf{q}, \mathbf{P}, t)$, which is called the generating function of the canonical transformations. Indeed, from the fact that Eqn.(3.16) and Eqn.(3.17) differ by a total derivative, we can obtain that

$$\frac{\partial}{\partial q_k} F(\mathbf{q}, \mathbf{P}, t) = p_k \tag{3.18a}$$

$$\frac{\partial}{\partial P_k} F(\mathbf{q}, \mathbf{P}, t) = Q_k \tag{3.18b}$$

$$\frac{\partial}{\partial t} F(\mathbf{q}, \mathbf{P}, t) + H = K. \tag{3.18c}$$

---

[5]For convenience the symbols $\mathbf{q}$ and $\mathbf{p}$ will be used to indicate respectively $q_1, ..., q_f$ and $p_1, ..., p_f$. In the same way for $Q_1, ..., Q_f$ and $P_1, ..., P_f$.

Choosing conveniently the canonical transformation, it is possible to obtain a transformation such that $K(Q_1, ..., Q_f, P_1, ..., P_f, t) = 0$ in Eqn.(3.18c). A canonical transformation of this kind is given by a $F(\mathbf{q}, \mathbf{P}, t)$ such that, using Eqn.(3.18c) and Eqn.(3.18a), it satisfies the equation:

$$\frac{\partial}{\partial t}F(\mathbf{q}, \mathbf{P}, t) + H(\mathbf{q}, \mathbf{p}, t) = 0 \Rightarrow \frac{\partial}{\partial t}F(\mathbf{q}, \mathbf{P}, t) + H\left(\mathbf{q}, \frac{\partial}{\partial \mathbf{q}}F(\mathbf{q}, \mathbf{P}, t), t\right) = 0.$$
(3.19)

This is the *Hamilton-Jacobi equation*: it is a nonlinear partial differential equation. It is possible to show that the solution of the Hamilton-Jacobi equation has the form:

$$F(\mathbf{q}, \mathbf{P}, t) = S_{cl}(\mathbf{q}, \mathbf{P}, t) = \int_{t_1}^{t} dt' \, L\left(\mathbf{q}_{cl}(t'), \dot{\mathbf{q}}_{cl}(t'), t'\right),$$
(3.20)

where the subscript cl indicates that the $\mathbf{q}(t)$ that we use is a classical solution and it satisfies the equation of motion with the boundary condition $\mathbf{q}_{cl}(t_1) = \mathbf{q}_1$ and $\mathbf{p}_{cl}(t_1) = \mathbf{P}$. Considering for simplicity the motion of a single particle of mass m in a potential $V(\mathbf{r})$, the Lagrangian in cartesian coordinates can be written as

$$L = \frac{m}{2}\dot{\mathbf{r}}^2 - V(\mathbf{r}) = \frac{\mathbf{p}^2}{2m} - V(\mathbf{r}),$$
(3.21)

from which, using Eqn.(3.18a), it follows[6]

$$\mathbf{p} = \frac{\partial L}{\partial \dot{\mathbf{r}}} = m\dot{\mathbf{r}} = \frac{\partial S_{cl}}{\partial \mathbf{r}} = \nabla S_{cl}.$$
(3.22)

Therefore the Hamilton-Jacobi equation becomes

$$\frac{\partial}{\partial t}S_{cl} + \frac{[\nabla S_{cl}]^2}{2m} + V = 0.$$
(3.23)

3.1.3 *WKB solutions*

Let us go now back to quantum mechanics and let us consider how to use the formalism above in order to get approximate solutions to the Schrödinger equation. The WKB method starts from the assumption that the potential

---

[6]Here and in Eqn.(3.19) the symbol of derivative with respect to a vector has been used, namely

$$\frac{\partial}{\partial \mathbf{q}} \equiv \nabla_{\mathbf{q}}$$

$$\frac{\partial}{\partial \mathbf{r}} \equiv \nabla_{\mathbf{r}} = \nabla.$$

In reality, $\mathbf{q}$ is not a vector in the usual sense, a vector with respect to the group of rotations in $\mathbb{R}^3$, but it is simply an object with n components.

varies very slowly as a function of the spatial coordinates. For simplicity the one-dimensional time independent Schrödinger equation will be considered:

$$\frac{d^2\psi(x)}{dx^2} + \frac{2m}{\hbar^2}[E - V(x)]\psi(x) = 0. \tag{3.24}$$

If $V = const.$ and $E > V$, Eqn.(3.24) has solutions of the form $Ae^{\pm ikx}$ with $k \equiv \sqrt{2m(E-V)}/\hbar$, which are oscillating wave functions with amplitude A and constant wavelength $\lambda = 2\pi/k$. So, if the potential varies slowly with x in comparison to λ, it is reasonable to suppose that the wave function remains an oscillating one but with a wavelength and amplitude slowly varying. In the same way, if $V = const.$ but $E < V$ then Eqn.(3.24) has solutions $Ae^{\pm\kappa x}$ with $\kappa \equiv \sqrt{2m(V-E)}/\hbar$, namely wave functions which are exponentials. Also in this case, making the previous assumptions, the solution will remain an exponential one but with amplitude and wavelength slowly varying. It is evident that this approach will give a good level of approximation only in certain part of the domain of the wave function. Note in fact that, where $E \approx V$, the quantities λ and $1/\kappa$ diverge to $+\infty$, so it is not correct anymore to assume $V(x)$ to be a slowly varying quantity. These critical points are called *turning points* and, classically, they represent the points where the particle changes the direction of motion. The mathematical treatment of the solutions in a neighborhood of these points is rather laborious and goes beyond the aims of our discussion. Far away from the turning points, we can test a solution like

$$\psi(x) = \exp\left[\frac{i}{\hbar}u(x)\right], \tag{3.25}$$

where the function[7] $u(x)$ is not linear in x anymore, differently from before when we treated the constant potentials. Differently from what we did before in Eqn.(3.2), the following parametrization for $u(x)$ will be used:

$$u(x) = S + \frac{\hbar}{i}\ln A. \tag{3.26}$$

Inserting Eqn. (3.25) into Eqn. (3.24), two coupled equations can be found. These equations are the unidimensional analog of Eqns.(3.6a) and (3.6b)

$$(S')^2 - 2m(E - V) = \hbar^2\frac{A''}{A} \tag{3.27a}$$

$$2A'S' + AS'' = 0. \tag{3.27b}$$

---

[7]Obviously the function $u(x)$ must be complex.

From Eqn.(3.27b) by integration it follows that

$$A = \frac{const.}{(S')^{1/2}};$$ (3.28)

substituting this expression into Eqn.(3.27a) we find

$$(S')^2 = 2m(E - V) + \hbar^2 \left[\frac{3}{4} \left(\frac{S''}{S'} \right)^2 - \frac{1}{2} \frac{S'''}{S'} \right],$$ (3.29)

which is equivalent to the Schrödinger equation (3.24). The WKB approximation consists in expanding S in powers of $\hbar^2$

$$S = S_0 + \hbar^2 S_1 + ...$$ (3.30)

The reason why the expansion is in $\hbar^2$ and not in $\hbar$ is that in Eqn.(3.29) we have $\hbar^2$ and not $\hbar$. Eqn.(3.25) can then be written as

$$\psi(x) = A \exp\left[\frac{i}{\hbar} S \right] = A \exp\left[\frac{i}{\hbar} (S_0 + \hbar^2 S_1 + ...) \right].$$ (3.31)

Inserting Eqn.(3.30) into Eqn.(3.29), at zeroth order in $\hbar^2$ we get

$$(S_0')^2 = 2m(E - V).$$ (3.32)

Via the following definitions

$$\lambda \equiv \frac{1}{k} = \frac{\hbar}{\sqrt{2m(E - V)}}, \text{ for } E > V$$ (3.33a)

$$\ell \equiv \frac{1}{\kappa} = \frac{\hbar}{\sqrt{2m(V - E)}}, \text{ for } E < V$$ (3.33b)

Eqn.(3.32) becomes

$$S_0' = \pm \frac{\hbar}{\lambda} = \pm \hbar k, \text{ for } E > V$$ (3.34a)

$$S_0' = \pm i \frac{\hbar}{\ell} = \pm i \hbar \kappa, \text{ for } E < V.$$ (3.34b)

So at zeroth order in $\hbar$ the WKB solution is:

$$\psi(x) \approx A \exp\left[\frac{i}{\hbar} S_0 \right],$$ (3.35)

and using Eqn.(3.28) we get

$$\psi(x) \approx \psi_{WKB}(x) = \frac{C}{\sqrt{k(x)}} \exp\left[\pm i \int k(x)\, dx \right], \text{ for } E > V$$ (3.36a)

$$\psi(x) \approx \psi_{WKB}(x) = \frac{C}{\sqrt{\kappa(x)}} \exp\left[\pm \int \kappa(x)\, dx \right], \text{ for } E < V.$$ (3.36b)

The general approximated solutions will be expressed as a combination of the two contributions above with $+$ and $-$ sign. It is interesting to stress that, in the region classically significant $(E > V)$, we get:

$$\rho \equiv |\psi(x)|^2 \approx \frac{|C|^2}{k(x)} \propto \frac{1}{p(x)}. \tag{3.37}$$

This equation shows that the probability density to find the particle at x is inversely proportional to the linear momentum $p(x)$, namely to its velocity, exactly as what we would expect in classical mechanics. When an approximation method is used, it is always necessary to specify the applicability conditions. It is not satisfactory to just say that the WKB method works well when the potentials are slowly varying. In order to be more quantitative, we can notice from Eqn.(3.31) that the $\hbar^2$ order correction gives an extra phase contribution: a term $\exp[i\hbar S_1]$ in the WKB solution. This factor can be neglected only if $\hbar S_1 \ll 1$. To evaluate this it is enough to insert Eqn.(3.30) into Eqn.(3.29) and equate the terms of order $\hbar^2$. With $E > V$, we get

$$\hbar S_1' = \pm \left[\frac{1}{4}\lambda'' - \frac{1}{8}\frac{(\lambda')^2}{\lambda}\right] \implies \hbar S_1 = \pm \left[\frac{1}{4}\lambda' - \frac{1}{8}\int \frac{(\lambda')^2}{\lambda}\,\mathrm{d}x\right]. \tag{3.38}$$

From this, we find that the applicability criterion for the WKB approximation is simply

$$\lambda'(x) \ll 1, \text{ for } E > V \tag{3.39a}$$

$$\ell'(x) \ll 1, \text{ for } E < V \tag{3.39b}$$

or, using the expression of λ as a function of V, we get

$$\frac{|m\hbar V'(x)|}{|2m\left[E - V(x)\right]|^{3/2}} \ll 1. \tag{3.40}$$

3.1.4 *Connection formulas*

When $E = V$, the quantities λ and ℓ diverge and the approximation is not valid anymore: note that the l.h.s. of Eqn.(3.40) diverges at the turning points too. Let us choose for example that $E \lessgtr V$ for $x \lessgtr a$. Then if the WKB can be applied everywhere except for the neighborhood of the turning point $x = a$, we have that the general solution is:

$$\psi_{WKB}(x) = \frac{A}{\sqrt{\kappa(x)}} \exp\left[-\int_a^x \kappa(x')\,\mathrm{d}x'\right]$$

$$+ \frac{B}{\sqrt{\kappa(x)}} \exp\left[+\int_a^x \kappa(x')\,\mathrm{d}x'\right], \text{ for } x < a \tag{3.41}$$

and

$$\psi_{WKB}(x) = \frac{C}{\sqrt{k(x)}} \exp\left[-i\int_a^x k(x')\,dx'\right]$$
$$+ \frac{D}{\sqrt{k(x)}} \exp\left[+i\int_a^x k(x')\,dx'\right], \text{ for } x > a \quad (3.42)$$

where A, B, C, D are four constants to be determined. The problem is to find out the relation between the coefficients A, B, C, D. If we were able to do that, we could neglect the exact solution of the Schrödinger equation in the neighborhood of the turning points where the WKB approximation fails. The relation among them can be found and we report here the so-called *connection formulas*, which allow us to connect the exponential WKB solutions to the oscillating ones

$$\frac{A}{\sqrt{\kappa(x)}} \exp\left[-\int_x^a \kappa(x')\,dx'\right] + \frac{B}{\sqrt{\kappa(x)}} \exp\left[+\int_x^a \kappa(x')\,dx'\right]$$
$$\longleftrightarrow \frac{2A}{\sqrt{k(x)}} \cos\left[\int_a^x k(x')\,dx' - \frac{\pi}{4}\right] - \frac{B}{\sqrt{k(x)}} \sin\left[\int_a^x k(x')\,dx' - \frac{\pi}{4}\right]$$
$$(3.43)$$

where, clearly, the exponentials are valid for $x \ll a$, while the trigonometric functions are valid for $x \gg a$. Analogous formulas hold for situations in which $V \lessgtr E$ for $x \lessgtr a$.

3.2   WKB in the path integral language

3.2.1   *Stationary phase method*

In this section we will strictly follow the book [Schulman (1981)]. Before treating in detail the WKB method via the path integral, we present some heuristic considerations about the so-called *stationary phase method*. Let us consider integrals of the form

$$F(\lambda) = \int_{-\infty}^{+\infty} dt\, e^{i\lambda f(t)} \quad (3.44)$$

and let us look for the dominant contributions in the limit $\lambda \to \infty$. The reason why we are interested in this problem is that we want to study the path integral in the limit $\hbar \to 0$. For great values of the parameter λ, the phase will vary rapidly and cancel out in the integration domain

$(t \in \mathbb{R})$, except where $f' = 0$. Therefore, the more important contributions will come from the neighborhood of the extremal points of the function f. Indeed, when the integration is performed over the regions where $f' \neq 0$, we have the sum of terms with very different relative phase and so, on average, there will be destructive interference that gives zero contribution to the integral. On the contrary, where the phase is stationary, at the first order there is no variation in f, therefore there is a non-zero contribution to the integral. Let's suppose that $f' = 0$ only for $t = t_0$. It follows

$$F(\lambda) = \int\limits_{-\infty}^{+\infty} dt \exp\left[i\lambda f(t_0) + \frac{1}{2} i\lambda f''(t_0)(t - t_0)^2 + ... \right]. \qquad (3.45)$$

Neglecting contributions higher than the second order, we get

$$F(\lambda) \approx e^{i\lambda f(t_0)} \int\limits_{-\infty}^{+\infty} dt \exp\left[-\frac{1}{2i} \lambda f''(t_0)(t - t_0)^2 \right] = e^{i\lambda f(t_0)} \sqrt{\frac{2i\pi}{\lambda f''(t_0)}}. \qquad (3.46)$$

In order to justify heuristically the previous approximation, let's consider the particular case with the form:

$$G(\lambda) = \int\limits_{-\infty}^{+\infty} dt \; e^{i\lambda t^2} e^{ia\lambda t^3} e^{ib\lambda t^4}. \qquad (3.47)$$

Taking the t expansion of the exponential with powers greater that t^2, we get:

$$G(\lambda) = \int\limits_{-\infty}^{+\infty} dt \; e^{i\lambda t^2} \left[1 + ia\lambda t^3 + ib\lambda t^4 - \frac{1}{2}a^2\lambda^2 t^6 + O\left(t^7\right) \right]$$

$$= \sqrt{\frac{i\pi}{\lambda}} \left(1 - \frac{3ib}{4\lambda} + \frac{15ia^2}{16\lambda} + ... \right). \qquad (3.48)$$

From the previous expansion we can see that the powers higher that t^2 give contributions that go to zero in the limit $\lambda \to \infty$, so it is as if each factor λt^2 would give a contribution of order 1, i.e.:

$$t = O\left(1/\sqrt{\lambda}\right). \qquad (3.49)$$

Hence, λt^3 is of order $1/\sqrt{\lambda}$ and so on. If there are more than one extremal points for the function f, the integral $F(\lambda)$ will be approximated by:

$$F(\lambda) \approx \sum_k e^{i\lambda f(t_k)} \sqrt{\frac{2i\pi}{\lambda f''(t_k)}}, \qquad (3.50)$$

where the sum is over all the extremal points of the function.

Let us now extend the idea above to the path integral. In the propagator

$$K(2,1) = \sum_{\substack{\text{all trajectories} \\ (a,t_1) \rightsquigarrow (b,t_2)}} e^{iS[x(t)]/\hbar} \tag{3.51}$$

in the limit $\hbar \to 0$, the most significant contribution will come from the trajectories for which the action is stationary:

$$\frac{\delta S[x(t)]}{\delta x} = 0, \tag{3.52}$$

i.e. exactly from the classical trajectories. The reader may ask what happens if there are more than one classical trajectory between the points 1 and 2 in the configuration space. This event can happen when the time interval $t_2 - t_1$ is large (see [Schulman (1981)]). In order to observe such events the difference of the action between the two trajectories must be comparable to $\hbar$. The propagator will then be

$$K(2,1) \approx A e^{iS_1/\hbar} + B e^{iS_2/\hbar}, \tag{3.53}$$

where S_1 and S_2 are the actions for the first and the second classical trajectory. The probability density would then be given by

$$|K(2,1)|^2 = \left| A e^{iS_1/\hbar} + B e^{iS_2/\hbar} \right|^2$$
$$= |A|^2 + |B|^2 + A B^* e^{i(S_1 - S_2)/\hbar} + A^* B e^{i(S_2 - S_1)/\hbar}. \tag{3.54}$$

From the previous expression we can see that the interference effect $2\text{Re}\left[A B^* e^{i(S_1 - S_2)/\hbar} \right]$ gives significant contributions only if $(S_1 - S_2) \sim \hbar$.

3.2.2   *Jacobi equation and Van Vleck determinant*

In the functional integrals for the limit $\hbar \to 0$ we should get a piece similar to the factor $\sqrt{2i\pi/\lambda f''(t_0)}$ of Eqn.(3.46). In order to evaluate this factor for the path integrals, we need to go back to classical mechanics. Let us consider (for a unidimensional system) the family of classical trajectories starting from the point $x = a$ at the time $t = 0$, without any indication of the arrival point. We can distinguish these trajectories from each other according to the value of the momentum P at $t = 0$ and let us indicate them as

$$x_{cl}(P,t). \tag{3.55}$$

We want to see how two close-by trajectories move away from each other. Let us consider first the quantity

$$J(P,t) \equiv \frac{\partial x_{cl}(P,t)}{\partial P}. \tag{3.56}$$

The difference between the two trajectories will then be:

$$x_{cl}(P+\varepsilon,t) - x_{cl}(P,t) = \varepsilon J(P,t) + O(\varepsilon^2). \tag{3.57}$$

We know that each $x_{cl}(P,t)$ satisfies the Euler-Lagrange equation:

$$\frac{d}{dt}\left(\frac{\partial L}{\partial \dot{x}_{cl}}\right) - \frac{\partial L}{\partial x_{cl}} = 0 \tag{3.58}$$

and since the family of classical trajectories depend on P, we can differentiate both sides of Eqn.(3.58) with respect to P

$$\frac{\partial}{\partial P} = \frac{\partial x_{cl}}{\partial P}\frac{\partial}{\partial x_{cl}} + \frac{\partial \dot{x}_{cl}}{\partial P}\frac{\partial}{\partial \dot{x}_{cl}} \tag{3.59}$$

and, furthermore, because P does not depend on time, we have that

$$\frac{\partial \dot{x}_{cl}}{\partial P} = \frac{\partial^2 x_{cl}}{\partial P \partial t} = \frac{d}{dt}\frac{\partial x_{cl}}{\partial P} = \dot{J}. \tag{3.60}$$

Eqn.(3.59) then becomes

$$\frac{\partial}{\partial P} = J\frac{\partial}{\partial x_{cl}} + \dot{J}\frac{\partial}{\partial \dot{x}_{cl}}. \tag{3.61}$$

Let us now evaluate the expressions:

$$\frac{\partial}{\partial P}\frac{d}{dt}\left(\frac{\partial L}{\partial \dot{x}_{cl}}\right) = \frac{d}{dt}\frac{\partial}{\partial P}\left(\frac{\partial L}{\partial \dot{x}_{cl}}\right) = \frac{d}{dt}\left(J\frac{\partial^2 L}{\partial x_{cl}\partial \dot{x}_{cl}} + \dot{J}\frac{\partial^2 L}{\partial \dot{x}_{cl}^2}\right) \tag{3.62a}$$

$$\frac{\partial}{\partial P}\frac{\partial L}{\partial x_{cl}} = J\frac{\partial^2 L}{\partial x_{cl}^2} + \dot{J}\frac{\partial^2 L}{\partial x_{cl}\partial \dot{x}_{cl}}. \tag{3.62b}$$

Eqn.(3.58) then becomes

$$\frac{d}{dt}\left(J\frac{\partial^2 L}{\partial x_{cl}\partial \dot{x}_{cl}} + \dot{J}\frac{\partial^2 L}{\partial \dot{x}_{cl}^2}\right) - J\frac{\partial^2 L}{\partial x_{cl}^2} - \dot{J}\frac{\partial^2 L}{\partial x_{cl}\partial \dot{x}_{cl}} = 0, \tag{3.63}$$

which, after simple calculations, gives

$$\frac{d}{dt}\left(\dot{J}\frac{\partial^2 L}{\partial \dot{x}_{cl}^2}\right) + \left[\frac{d}{dt}\left(\frac{\partial^2 L}{\partial x_{cl}\partial \dot{x}_{cl}}\right) - \frac{\partial^2 L}{\partial x_{cl}^2}\right]J = 0. \tag{3.64}$$

The previous expression is known as the Jacobi equation and its solution is called *Jacobi field.* If we use a Lagrangian of the form:

$$L = \frac{m}{2}\dot{x}^2 - V(x,t), \tag{3.65}$$

it follows that

$$\frac{\partial^2 L}{\partial x^2} = -\frac{\partial^2 V}{\partial x^2} \tag{3.66a}$$

$$\frac{\partial^2 L}{\partial x \partial \dot{x}} = 0 \tag{3.66b}$$

$$\frac{\partial^2 L}{\partial \dot{x}^2} = m. \tag{3.66c}$$

Hence the Jacobi equation simply becomes

$$m\frac{\partial^2 J}{\partial t^2} + \frac{\partial^2 V}{\partial x_{cl}^2}J = 0. \tag{3.67}$$

Since for this Lagrangian $p = m\dot{x}_{cl}$, from Eqn.(3.60) we find

$$\frac{\partial \dot{x}_{cl}(P,0)}{\partial P} = \frac{1}{m} \Rightarrow \frac{\partial J(P,0)}{\partial t} = \frac{1}{m}. \tag{3.68}$$

Moreover, because $x_{cl}(P,0) = a$, $\forall P$, using Eqn.(3.57) the following relation holds

$$J(P,0) = 0. \tag{3.69}$$

So, we have that the Jacobi field satisfies the following Cauchy problem:

$$\begin{cases} m\dfrac{\partial^2 J(P,t)}{\partial t^2} + \dfrac{\partial^2 V}{\partial x_{cl}^2}J(P,t) = 0 \\[4mm] J(P,0) = 0 \text{ and } \dfrac{\partial J(P,0)}{\partial t} = \dfrac{1}{m}. \end{cases} \tag{3.70}$$

As the previous equation is a second order one, we could have given boundary conditions like $J(P,0) = 0$ and $J(P,T) = b$. Among the boundary conditions the ones for which

$$J(P,T) = 0 \tag{3.71}$$

are particularly important. The point at $t = T$ which satisfies the above equation is called *conjugate point* or *focal point*. In Fig. 3.1 there is a family of classical trajectories which start from the same point at $t = 0$, diverge and ultimately converge at $t = T$. The possibility to have more classical trajectories with the same initial and final points was discussed by [Schulman (1981)] for certain kind of physical systems and for large time intervals. Note that different trajectories with the same initial and final points can exist only if we work in configuration space. If on the contrary we work in phase space and a fixed initial point and momentum are given, then only one classical trajectory is allowed.

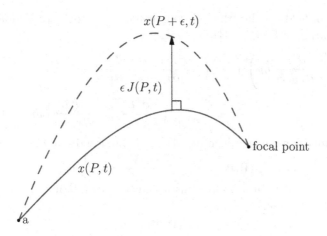

Fig. 3.1   Jacobi field representation.

Let $x_{cl}(P,T) = b$ and let us consider a classical trajectory connecting the points $(a,0)$ and (b,T). Changing a little the position of the end points a small change of the trajectory is expected. Using Eqns.(3.20) and (A.20) expressed in the variables (x_{cl}, a, t), and Eqn.(A.30) of Appendix A, we have that

$$p(0) = P = -\frac{\partial S_{cl}(x_{cl}, a, t)}{\partial a}. \tag{3.72}$$

P depends on the coordinates of the focal point (b,T), since it is calculated starting from the extremal trajectory x_{cl}, which itself depends on b. Therefore we can differentiate this last equation with respect to b

$$\frac{\partial P}{\partial b} = -\frac{\partial^2 S_{cl}}{\partial a \partial b} = -\frac{\partial^2 S_{cl}(x_{cl}, a, t)}{\partial a \partial x_{cl}}\bigg|_{x_{cl}=b} = \frac{1}{J}. \tag{3.73}$$

In general, working with higher dimensional systems (n for example), the Jacobi field becomes a matrix $\underline{\underline{J}}$ of dimensions $n \times n$, whose definition is the following[8]

$$J_{ik}(\mathbf{P}, t) \equiv \frac{\partial x_i(\mathbf{P}, t)}{\partial P_k}, \tag{3.74}$$

where $x_i(\mathbf{P}, 0) = a_i$, $\forall \mathbf{P}$. Using the vector representation the previous definition becomes

$$\underline{\underline{J}}(\mathbf{P}, t) \equiv \frac{\partial \mathbf{x}(\mathbf{P}, t)}{\partial \mathbf{P}}. \tag{3.75}$$

---

[8]From now on, we will omit the subscript (cl) for the extremal classical trajectories.

The Jacobi field satisfies the following equations, which are n dimensional generalization of Eqn. (3.64) :

$$\frac{d}{dt}\left(\sum_{l=1}^{n}\frac{\partial^2 L}{\partial \dot{x}_i \partial \dot{x}_l}\dot{J}_{lk}\right) + \sum_{l=1}^{n}\left(\frac{\partial^2 L}{\partial \dot{x}_i \partial x_l} - \frac{\partial^2 L}{\partial x_i \partial \dot{x}_l}\right)\dot{J}_{lk}$$

$$+ \sum_{l=1}^{n}\left[\frac{d}{dt}\left(\frac{\partial^2 L}{\partial \dot{x}_i \partial x_l}\right) - \frac{\partial^2 L}{\partial x_i \partial x_l}\right]J_{lk} = 0 \quad (3.76)$$

and n dimensional generalizations of Eqns.(3.68) and (3.69):

$$J_{ik}\left(\mathbf{P},0\right) = 0 \qquad \frac{\partial J_{ik}\left(\mathbf{P},0\right)}{\partial t} = \frac{1}{m}\delta_{ik}, \qquad (3.77)$$

with $i,k = 1,...,n$. In n dimensions it can be shown that if at $t = T$ there is a focal point, then

$$\det \underline{\underline{\mathbf{J}}}\left(\mathbf{P},T\right) = 0\,. \qquad (3.78)$$

In more dimensions Eqn.(3.73) will be a matrix whose determinant, called *Van Vleck determinant*, is defined as follows

$$D \equiv \det\left\{-\frac{\partial^2 S_{cl}}{\partial a_i \partial b_k}\right\} = \det\left(-\frac{\partial^2 S_{cl}}{\partial \mathbf{a}\partial \mathbf{b}}\right). \qquad (3.79)$$

From Eqn.(3.73), it can be proved that this determinant is the inverse of the determinant of the Jacobi field matrix. In order to have a more physical interpretation of D it can be shown that it satisfies the following continuity equation

$$\sum_{k=1}^{n}\frac{\partial}{\partial b_k}\left(v_k D\right) + \frac{\partial D}{\partial t} = 0, \qquad (3.80)$$

where v_k is the velocity at time T. Therefore, remembering the hydrodynamic continuity equation for the density of an incompressible fluid, we can say that D represents a *density of trajectories* . At the focal points we get that $D \to \infty$ because of Eqn.(3.78) and then, according to the previous physical interpretation, we have an infinite density of trajectories since all the paths converge to these points.

3.3 The semiclassical propagator

Using what we have obtained in the previous sections, we can now find out the explicit expression of the semiclassical approximation for the propagator $K\left(2,1\right)$. From the analysis of the stationary phase method, the WKB approximation consists in replacing an arbitrary Lagrangian with a quadratic one to which the results obtained for the quadratic potentials can be applied.

3.3.1 Steady phase approximation method for the path integral

Following [Schulman (1981)], we start from Eqn.(2.12) for the propagator, with the discretized action :

$$S_N = \varepsilon \sum_{j=1}^{N} \left[\frac{m}{2} \left(\frac{x_j - x_{j-1}}{\varepsilon} \right)^2 - V(x_j, t_j) \right], \qquad (3.81)$$

where (see Fig. 2.1), $t_0 = 0$, $t_N = t$, $x_N = x$ and $\varepsilon = t/N$. For fixed and finite N, let us apply the steady phase approximation to the variables $x_1, ..., x_{N-1}$. The extremal points of Eqn.(3.81) can be found by imposing

$$\frac{\partial S_N}{\partial x_j} = 0 \qquad (j = 1, ..., N-1), \qquad (3.82)$$

that is

$$m \frac{x_{j+1} - 2x_j + x_{j-1}}{\varepsilon^2} = -\frac{\partial V(x_j, t_j)}{\partial x_j} \qquad (j = 1, ..., N-1). \qquad (3.83)$$

These are the discretized version of the equations of motion. If we indicate with $x_{cl,j}$ $(j = 1, ..., N-1)$ their solutions, we can proceed as it was done before in the study of the quadratic Lagrangian, expanding the action S_N around this solution. The only difference consists in the discretized form of the functions and differential operators entering the action. Let us use the notation

$$\eta_j \equiv x_j - x_{cl,j} \quad \text{with } \eta_0 = \eta_N = 0. \qquad (3.84)$$

This expansion is analogous to Eqn.(2.23), with the difference that, in general, we will have contributions different from zero at all orders[9]:

$$S_N[x] = S_N[x_{cl} + \eta] = S_N[x_{cl}] + \sum_{k=1}^{\infty} \frac{\delta^k S_N}{\delta x^k} \bigg|_{x=x_{cl}} \frac{\eta^k}{k!}$$

$$= S_N[x_{cl}] + \delta S_N[x_{cl}] + \frac{1}{2} \delta^2 S_N[x_{cl}] + \sum_{k=3}^{\infty} \frac{1}{k!} \delta^k S_N[x_{cl}]. \qquad (3.85)$$

Now it is evident the importance of the steady phase approximation, since it gives us the possibility, by using the classical trajectory, to eliminate in Eqn.(3.85) not only the first variation but also all the variations of the action beyond the third order, so

$$S_N[x] \approx S_N[x_{cl}] + \frac{1}{2} \delta^2 S_N[x_{cl}]. \qquad (3.86)$$

---

[9]We will indicate with x the collection of all x_j, namely the discretization of $x(t)$. In the same way we will proceed for $x_{cl}(t)$ and $\eta(t)$.

We now want to justify this approximation. Let us recall the considerations we made regarding the integral (3.47). Let us suppose, in the same way, that

$$\frac{\delta^2 S}{\hbar} = O(1) \quad \text{for } \hbar \to 0. \tag{3.87}$$

Remembering that the second variation depends quadratically only on $\eta = x - x_{cl}$, we can say that

$$\eta = O\left(\sqrt{\hbar}\right) \quad \text{for } \hbar \to 0. \tag{3.88}$$

Therefore, the third variation $\delta^3 S$, which goes as η^3, will have an additional power $\sqrt{\hbar}$ and so

$$\frac{\delta^3 S}{\hbar} = O\left(\sqrt{\hbar}\right) \quad \text{for } \hbar \to 0. \tag{3.89}$$

Since higher order variations will be consequently smaller, at first order in $\hbar$ only $\delta^2 S$ contributes to the path integral.

3.3.2 *Approximated path integral evaluation*

Starting from Eqn.(3.86), it can also be shown that in the discretized case the first variation is equal to zero. Then, the only contribution to take into account is the second variation for which the following expression can be derived (see Appendix C):

$$\frac{1}{2}\delta^2 S_N[x_{cl}] = \frac{m}{2}\varepsilon \sum_{k=0}^{N} \left[\left(\frac{\eta_{\alpha_{k+1}} - \eta_{\alpha_k}}{\varepsilon}\right)^2 - \frac{1}{m}\frac{\partial^2 V}{\partial x_{cl,\alpha_k}^2}\eta_{\alpha_k}^2\right]. \tag{3.90}$$

First of all let us notice the strict formal analogy of Eqn.(3.90) with the discretized action (2.34). We have basically a harmonic oscillator with frequency $\Omega(t)$ depending on time, whose expression is

$$\Omega^2(t) \equiv \frac{1}{m}\frac{\partial^2 V(x(t),t)}{\partial x(t)^2}\bigg|_{x(t)=x_{cl}(t)}. \tag{3.91}$$

We can go ahead by defining the vector η as in Eqn. (2.35). As for the matrix $\underline{\sigma}$, let us first define

$$\Omega_k^2 \equiv \frac{1}{m}\frac{\partial^2 V}{\partial x_{cl,\alpha_k}^2}. \tag{3.92}$$

With respect to Section 2.6, here we will have some different factors. Factorizing ε we have:

$$\underline{\sigma} \equiv \underline{\underline{A}} + \underline{\underline{B}}, \tag{3.93}$$

where

$$\underline{\underline{A}} \equiv \frac{m}{2}\frac{1}{\varepsilon^2}\begin{pmatrix} 2 & -1 & & & 0 \\ -1 & 2 & -1 & & \\ & \ddots & \ddots & \ddots & \\ & & -1 & 2 & -1 \\ 0 & & & -1 & 2 \end{pmatrix} \quad \text{and} \quad \underline{\underline{B}} \equiv -\frac{m}{2}\begin{pmatrix} \Omega_1^2 & & & 0 \\ & \Omega_2^2 & & \\ & & \ddots & \\ & & & \Omega_{N-1}^2 \\ 0 & & & & \Omega_N^2 \end{pmatrix}.$$

$$(3.94)$$

So the second variation can be written as

$$\frac{1}{2}\delta^2 S_N[x_{cl}] = \varepsilon \eta^T \underline{\underline{\sigma}} \eta. \tag{3.95}$$

We want to diagonalize $\underline{\underline{\sigma}}$, which means we have to solve the following eigenvalue problem

$$\underline{\underline{\sigma}}\chi = \lambda\chi. \tag{3.96}$$

Since the matrix $\underline{\underline{\sigma}}$ is symmetric, it will have N eigenvalues λ_k and N eigenvectors χ_k. Note that the matrix $\underline{\underline{A}}$ above is a Jacobi matrix and the expression for its eigenvalues ℓ_k and eigenvectors ξ_k is well known (see [Schulman (1981)]). Therefore the matrix $\underline{\underline{\sigma}}$ can be considered a Jacobi matrix only if the matrix $\underline{\underline{B}}$, containing the Ω_k^2, can be neglected. So, taking $\underline{\underline{\sigma}} \approx \underline{\underline{A}}$, we have

$$\lambda_k = \frac{2m}{\varepsilon^2}\sin^2\left(\frac{k}{2}\frac{\pi}{N+1}\right) + \begin{pmatrix} \text{small terms coming} \\ \text{from } \underline{\underline{B}} \end{pmatrix} \quad (k = 1, ..., N). \tag{3.97}$$

We see that for $N \to \infty$ and k small compared to N we have that

$$\lambda_k \sim \frac{m}{2T^2}\pi^2 k^2, \tag{3.98}$$

which is a behavior that recalls the unidimensional infinite potential wall.

Going back to the second variation, let us use the diagonalization of $\underline{\underline{\sigma}}$. If the following unitary matrix,

$$\underline{\underline{U}} \equiv \begin{pmatrix} (\chi_1)_1 & (\chi_2)_1 & \cdots & (\chi_N)_1 \\ (\chi_1)_2 & (\chi_2)_2 & \cdots & (\chi_N)_2 \\ \vdots & \vdots & & \vdots \\ (\chi_1)_N & (\chi_2)_N & \cdots & (\chi_N)_N \end{pmatrix}, \tag{3.99}$$

which has as columns the normalized eigenvectors, is considered, then it can be shown[10] that $\underline{\underline{U}}^T\underline{\underline{U}} = \underline{\underline{U}}\,\underline{\underline{U}}^T = \mathbb{I}/\varepsilon$. In the limit $N \to \infty$ the

---

[10]The presence of the ε factor is crucial in order not to have divergences in the limit $N \to \infty$.

eigenvectors represent a complete orthonormal set $\{\chi_k\}$, over which η can be expanded :

$$\eta_{\alpha_i} = \sum_{k=1}^{N} \zeta_k \left(\chi_k\right)_i \quad \text{or} \quad \eta = \underline{\underline{U}} \zeta, \tag{3.100}$$

where ζ is the vector of the coefficients entering the expansion. Using the previous expression the second variation can be written as:

$$\frac{1}{2}\delta^2 S_N[x_{cl}] = \varepsilon \eta^T \underline{\underline{\sigma}} \eta = \varepsilon \zeta^T \underline{\underline{U}}^T \underline{\underline{\sigma}} \, \underline{\underline{U}} \zeta = \sum_{k=1}^{N} \lambda_k \zeta_k^2. \tag{3.101}$$

Next, the propagator integrations have to be changed as in Eqn.(2.38). Note that here we have $\left|\det \underline{\underline{U}}\right| = \sqrt{\det \underline{\underline{U}}^T \underline{\underline{U}}} = \varepsilon^{-N/2}$ and exactly for the same reasons explained in the previous footnote. Without repeating the steps done before, we can conclude that the semiclassical propagator is

$$K_{WKB}\left(b,T;a,0\right) = \lim_{N\to\infty} \left(\frac{m}{2\pi\hbar i}\right)^{(N+1)/2} \varepsilon^{-N-1/2} \exp\left(\frac{i}{\hbar}S_N[x_{cl}]\right)$$

$$\times \int_{-\infty}^{+\infty} d\zeta_1 \int_{-\infty}^{+\infty} d\zeta_2 ... \int_{-\infty}^{+\infty} d\zeta_N \exp\left(\frac{i}{\hbar}\sum_{k=1}^{N}\lambda_k\zeta_k^2\right). \tag{3.102}$$

Performing the N gaussian integrations, we obtain

$$K_{WKB}\left(b,T;a,0\right) = \lim_{N\to\infty} \left(\frac{m}{2\pi\hbar i}\right)^{(N+1)/2} \varepsilon^{-N-1/2}$$

$$\times \exp\left(\frac{i}{\hbar}S_N[x_{cl}]\right) \prod_{k=1}^{N} \sqrt{\frac{\pi i \hbar}{\lambda_k}}. \tag{3.103}$$

From Section 2.6 on quadratic Lagrangian we get that, for $N \to \infty$,

$$K_{WKB}\left(b,T;a,0\right) = \sqrt{\frac{m}{2\pi\hbar i f\left(T,0\right)}} \exp\left(\frac{i}{\hbar}S_{cl}\right), \tag{3.104}$$

where f satisfies the following relations (with $t_1 = 0$ and $t_2 = T$)

$$\begin{cases} \dfrac{\partial^2 f\left(t_2,t_1\right)}{\partial t_2^2} + \Omega^2\left(t_2\right) f\left(t_2,t_1\right) = 0 \\[3mm] f\left(t_1,t_1\right) = 0 \ \text{and} \ \dfrac{\partial f\left(t_2,t_1\right)}{\partial t_2}\bigg|_{t_2=t_1} = 1. \end{cases} \tag{3.105}$$

3.3.3 Functional determinants

In this section we will do a short excursus that will help us understand in a deeper way what we did before. Following [Kleinert (1990)], let us consider the continuum version of the discretized expression (3.90)

$$\frac{1}{2}\delta^2 S[x_{cl}(t)] = \int_{t_1}^{t_2} dt \frac{m}{2} \left[\left(\frac{d\eta(t)}{dt}\right)^2 - \Omega^2(t)\eta^2(t) \right], \qquad (3.106)$$

where $x_{cl}(t)$ is the solution of the equations of motion with boundary conditions $x_{cl}(t_1 = 0) = a$ and $x_{cl}(t_2 = T) = b$ and where $\eta(t)$ is a *fluctuation*, namely a real function such that $\eta(t_1 = 0) = \eta(t_2 = T) = 0$. Integrating by parts, Eqn. (3.106) becomes

$$\frac{1}{2}\delta^2 S[x_{cl}(t)] = \frac{m}{2}\eta(t)\frac{d\eta(t)}{dt}\bigg|_{t_1}^{t_2} + \int_{t_1}^{t_2} dt \frac{m}{2}\left[-\eta(t)\frac{d^2\eta(t)}{dt^2} - \Omega^2(t)\eta^2(t)\right]$$

$$= \frac{m}{2}\int_{t_1}^{t_2} dt\, \eta(t)\left[-\frac{d^2}{dt^2} - \Omega^2(t)\right]\eta(t). \qquad (3.107)$$

This is the continuum version of Eqn. (C.13) presented in Appendix C. According to these considerations, it is easy to see that $\underline{\sigma}$ (modulo the multiplicative constant $m/2$) is the discretization of the differential operator

$$\hat{M} \equiv -\partial_t^2 - \Omega^2(t) \qquad (3.108)$$

and the calculation of $\det \underline{\sigma}$ consists of the evaluation of the functional determinant of the operator $\det\left[-\partial_t^2 - \Omega^2(t)\right]$. The determinant of an operator is given by the product of its eigenvalues. Therefore we have to get the eigenvalues of $\hat{M}$ by solving the following equation[11]

$$-\frac{d^2\chi_k(t)}{dt^2} - \Omega^2(t)\chi_k(t) = \lambda_k\chi_k(t), \qquad (3.109)$$

with the conditions $\chi_k(t_1 = 0) = \chi_k(t_2 = T) = 0$, $\forall k$. The previous one is a Schrödinger-like equation for a unidimensional system (in coordinate t) with a potential $-\Omega^2(t)$. Indeed, the class of potentials for which this equation can be solved is well known. Supposing that the eigenfunctions $\{\chi_k(t)\}$ are a complete orthonormal set in the interval $[t_1, t_2]$, namely

$$\int_{t_1}^{t_2} dt\, \chi_k^*(t)\chi_j(t) = \delta_{kj} \qquad (3.110a)$$

$$\sum_k \chi_k^*(t)\chi_k(t') = \delta(t - t'), \qquad (3.110b)$$

---

[11]As $\hat{M}$ is a Hermitian operator it has real eigenvalues λ_k.

then the *fluctuation* $\eta(t)$ can be expanded over $\{\chi_k(t)\}$ with coefficients ζ_k, i.e.:

$$\eta(t) = \sum_k \zeta_k \chi_k(t). \tag{3.111}$$

Inserting Eqn.(3.111) into Eqn.(3.107) and making use of the properties of the eigenfunctions we find that

$$\frac{1}{2}\delta^2 S[x_{cl}(t)] = \frac{m}{2} \sum_k \sum_j \zeta_k \zeta_j \int_{t_1}^{t_2} dt\, \chi_k(t) \left[-\frac{d^2}{dt^2} - \Omega^2(t) \right] \chi_j(t)$$

$$= \frac{m}{2} \sum_k \sum_j \zeta_k \zeta_j \lambda_k \int_{t_1}^{t_2} dt\, \chi_k(t)\chi_j(t)$$

$$= \frac{m}{2} \sum_k \sum_j \zeta_k \zeta_j \lambda_k \delta_{kj} = \frac{m}{2} \sum_k \lambda_k \zeta_k^2. \tag{3.112}$$

This reminds us of Eqn.(3.101). At this point, it is possible to proceed as before by performing the gaussian integrations in the path integral. Also in this case we shall obtain $\prod \lambda_k$, that is $\det \hat{M}$. Indeed, working in the continuum limit, the evaluation of the remaining normalization factors present in the path integral turns out to be more laborious and requires specific theorems on functional determinants. Working in the discretized form, we have been able to evaluate the function $f(t_2,t_1)$, introduced in Eqn.(2.40), that enters the propagator $K_{WKB}(b,t_2;a,t_1)$ as a normalization factor in front of the phase $\exp(iS_{cl}/\hbar)$. It is the kernel of the operator $\left[-\partial_{t_2}^2 - \Omega^2(t_2) \right]$:

$$\left[-\partial_{t_2}^2 - \Omega^2(t_2) \right] f(t_2,t_1) = 0. \tag{3.113}$$

More sophisticated considerations, done in the continuum limit (see [Coleman (1985)]), lead to the same results.

3.3.4 *Final expression*

Let us go now back to the derivation of Section 3.3.2. We will generalize expression (3.104) to the case of higher dimensions and we will build up the functional determinant of Van Vleck. Observe, first, that $f(t_2,t_1)$ of Eqn.(3.104) satisfies a Cauchy problem (3.105), that is the same we have encountered before for the Jacobi field $J(P,t)$ (that is Eqn. (3.70)) modulo initial condition. If the following function is defined,

$$\tilde{f}(t_2,t_1) \equiv \frac{f(t_2,t_1)}{m}, \tag{3.114}$$

it is easy to see that it satisfies the same Cauchy problem as $J(P,t)$. Therefore we get

$$\tilde{f}(t_2, t_1) = J(P,t). \tag{3.115}$$

Starting from Eqn.(3.104) and recalling Eqn.(3.73), we find

$$K_{WKB}(b, T; a, 0) = \sqrt{\frac{m}{2\pi\hbar i f(T,0)}} \exp\left(\frac{i}{\hbar}S_{cl}\right) = \sqrt{\frac{m}{2\pi\hbar i m J}} \exp\left(\frac{i}{\hbar}S_{cl}\right)$$

$$= \sqrt{\frac{i}{2\pi\hbar} \frac{\partial^2 S_{cl}}{\partial a \partial b}} \exp\left(\frac{i}{\hbar}S_{cl}\right). \tag{3.116}$$

The n dimensional generalization of Eqn.(3.116) is

$$K_{WKB}(\mathbf{b}, t_2; \mathbf{a}, t_1) = \sqrt{\det\left(\frac{i}{2\pi\hbar} \frac{\partial^2 S_{cl}}{\partial \mathbf{a} \partial \mathbf{b}}\right)} \exp\left(\frac{i}{\hbar}S_{cl}\right), \tag{3.117}$$

where, as before, the action function is evaluated along an extremal classical trajectory $\mathbf{x}_{cl}(t)$ connecting the points $(\mathbf{a}, t_1)$ and $(\mathbf{b}, t_2)$, namely

$$S_{cl} = \int_{t_1}^{t_2} dt' \, L\left(\mathbf{x}_{cl}(t'), \dot{\mathbf{x}}_{cl}(t'), t'\right). \tag{3.118}$$

Recalling the definition (3.79), from Eqn.(3.117) we get

$$K_{WKB}(\mathbf{b}, t_2; \mathbf{a}, t_1) = \sqrt{\left(\frac{1}{2\pi i\hbar}\right)^n \det\left(-\frac{\partial^2 S_{cl}}{\partial \mathbf{a} \partial \mathbf{b}}\right)} \exp\left(\frac{i}{\hbar}S_{cl}\right) \tag{3.119}$$

$$= \left(\frac{1}{2\pi i\hbar}\right)^{n/2} \sqrt{D} \exp\left(\frac{i}{\hbar}S_{cl}\right), \tag{3.120}$$

which contains the Van Vleck determinant D.

Using the fact that D is the inverse of the determinant of the Jacobi field matrix (see Section 3.2.2) we have that

$$K_{WKB}(\mathbf{b}, t_2; \mathbf{a}, t_1) = \left(\frac{1}{2\pi i\hbar}\right)^{n/2} \frac{1}{\sqrt{\det \underline{\mathbf{J}}}} \exp\left(\frac{i}{\hbar}S_{cl}\right). \tag{3.121}$$

We notice that the volume of the propagator at the focal points diverges because at those points we have :

$$\det \underline{\mathbf{J}} = 0. \tag{3.122}$$

These points are also called *caustics* and the word comes from optics. These divergences are present because we are calculating the propagator in an approximated way. If we were able to calculate it exactly there would be no divergences. From what we have found, we can say that the WKB method gives *exact* results for quadratic Lagrangian for the obvious reason that the third variation is zero.

Chapter 4

Wigner Functions and its associated Path Integral

Weyl, Wigner and Mojal [Weyl (1927); Wigner (1932); Moyal (1949)] developed a formalism for quantum mechanics which is entirely based on *commuting* functions in phase space and not on operators. The correspondence between the operators and the functions is called *"symbol map"* and we are now going to define it.

Suppose the classical system lives in a phase space $\mathcal{M}$ on which complex-valued functions $A, B, C,...$ are defined:

$$A, B, C, \cdots \in Fun(\mathcal{M}) \equiv \mathcal{F}(\mathcal{M}) .$$

At the quantum level the system lives on some Hilbert space $\mathcal{V}$ and has some observables which we will indicate with $\widehat{A}, \widehat{B}, \widehat{C},....$ The *"symbol map"* associates a unique function in $\mathcal{M}$ to each operator $\widehat{A}, \widehat{B}, \widehat{C},...$:

$$A = symb(\widehat{A})$$
$$B = symb(\widehat{B})$$
$$C = symb(\widehat{C}).$$

The non-commutative nature of QM is reflected in the fact that on $\mathcal{F}(\mathcal{M})$ we will define a non-commutative product called *"star product"*, indicated by the symbol $*$. It acts in the following manner:

$$symb(\widehat{A}\,\widehat{B}) = symb(\widehat{A}) * symb(\widehat{B}).$$

It is associative

$$A * (B * C) = (A * B) * C$$

but non-commutative

$$A * B \neq B * A.$$

49

It have shown rigorously [Bayen *et al.* (1978a,b)] that the star product is a
"*deformation*" of the ordinary pointwise multiplication among functions.

Let us assume that the Hilbert space $\mathcal{V}$ is the space of states of an arbitrary quantum mechanical system with N degrees of freedom and that the manifold $\mathcal{M} = \mathcal{M}_{2N}$ is the $2N$ dimensional phase space of the associated classical system. We will indicate the local coordinates on $\mathcal{M}_{2N}$ with $\varphi^a = (q^1, ..., q^N; p^1, ..., p^N)$ where $a = 1, ...2N,$. For simplicity we will assume that these coordinates can be extended globally. The phase space has a symplectic two form ω defined as:

$$\omega = \frac{1}{2}\, \omega_{ab}\, d\varphi^a \wedge d\varphi^b.$$

Locally ω_{ab} can be written as

$$\begin{pmatrix} 0 & -\mathbb{I}_N \\ \mathbb{I}_N & 0 \end{pmatrix}$$

and its inverse is

$$\begin{pmatrix} 0 & \mathbb{I}_N \\ -\mathbb{I}_N & 0 \end{pmatrix}.$$

We have used the opposite sign convention compared to [Abraham and Marsden (1978)]). Using ω^{ab} the usual Poisson brackets among two functions $A, B \in \mathcal{F}(\mathcal{M})$ can be written as

$$\{A, B\}_{pb}(\varphi) \equiv \partial_a A(\varphi)\, \omega^{ab}\, \partial_b B(\varphi)$$

where $\partial_a \equiv \dfrac{\partial}{\partial \varphi^a}$.

There are various possibilities of associating an operator $\widehat{A}(\hat{q}, \hat{p})$ to the function $A(q, p)$ via the inverse symbol map

$$\widehat{A}(\hat{q}, \hat{p}) = symb^{-1}\left(A(q, p)\right).$$

Typically different definitions of the symbol map correspond to different operator ordering prescriptions [Berezin (1981)]. In the following we shall mainly work with what is called the Weyl symbol [Berezin (1981)], which has the property that if $A(q, p)$ is a polynomial in p and q, the associated operator $\widehat{A}(\hat{q}, \hat{p})$ is the symmetrically ordered polynomial in $\hat{q}$ and $\hat{p}$, for example

$$symb^{-1}(p\,q) = \frac{1}{2}\left(\hat{p}\,\hat{q} + \hat{q}\,\hat{p}\right).$$

Let us now define the following operator $\hat{T}$

$$\hat{T}(\varphi_0^a) \equiv \exp\left[\frac{i}{\hbar}\hat{\varphi}^a\,\omega_{ab}\,\varphi_0^b\right]$$

$$= \exp\left[\frac{i}{\hbar}(p_0\,\hat{q} - q_0\,\hat{p})\right] \qquad (4.1)$$

where φ_0^a is a fixed point in phase space. The Weyl "*symbol*" associated to the operator $\hat{A}$ is defined as:

$$A(\varphi^a) = \int \frac{\mathrm{d}^{2N}\varphi_0^a}{(2\pi\hbar)^N}\,\exp\left[\frac{i}{\hbar}\varphi_0^a\,\omega_{ab}\,\varphi^b\right]\cdot\left[\hat{T}(\varphi_0^a)\,\hat{A}\right]. \qquad (4.2)$$

The inverse map reads as:

$$\hat{A} = \int \frac{\mathrm{d}^{2N}\varphi\,\mathrm{d}^{2N}\varphi_0}{(2\pi\hbar)^{2N}}\,A(\varphi)\,\exp\left[\frac{i}{\hbar}\varphi^a\,\omega_{ab}\,\varphi_0^b\right]\hat{T}(\varphi_0^a). \qquad (4.3)$$

This last expression is due to [Weyl (1927)]. It can be proved that the operators $\hat{T}(\varphi_0)$ form a complete and orthogonal (with respect to the Hilbert-Schmidt inner product) set of operators in terms of which any operator $\hat{A}$ can be expanded [Littlejohn (1986)]. In QM a particularly important class of operators are the density operators $\hat{\rho}$. Their symbol can be built via formula (4.2). In particular for pure states for which $\hat{\rho} = |\psi\rangle\langle\psi|$ one obtains from Eqn.(4.2):

$$\rho = \int \mathrm{d}^N x\,\exp\left[-\frac{ipx}{\hbar}\right]\,\psi\left(q + \frac{1}{2}x\right)\,\psi^\star\left(q - \frac{1}{2}x\right)$$

which is called *Wigner function* [Wigner (1932)]. The symbol $\rho(\varphi^a)$ is the *quantum* mechanical analogue of the classical probability density $\rho_{cl}(\varphi^a)$ used in *classical* statistical mechanics. However, different from classical mechanics, the *quantum* symbol $\rho(\psi)$ is not positive-definite and is often referred as a "*pseudo-probability density*". The usual positive-definite quantum mechanical distribution over position and momentum are recovered as:

$$|\psi(q)|^2 = \int \frac{\mathrm{d}^N p}{(2\pi\hbar)^N}\,\rho(q,p)$$

$$|\psi(p)|^2 = \int \frac{\mathrm{d}^N q}{(2\pi\hbar)^N}\,\rho(q,p).$$

It is also possible to prove that, given an operator $\hat{O}$ and a state $|\psi\rangle$, its expectation value can be written in terms of the Wigner function $\rho(p,q)$ associated to $|\psi\rangle$ and the symbol $O(\varphi)$ associated to the operator $\hat{O}$ in the following manner:

$$\langle\psi|\hat{O}|\psi\rangle = \int \frac{\mathrm{d}^{2N}\varphi}{(2\pi\hbar)^N}\,\rho(\varphi)\,O(\varphi). \qquad (4.4)$$

So "somehow" the quantum averages can be written in a "*classically-looking*" manner involving only C-number functions on $\mathcal{M}_{2N}$. The star product "$*$" mentioned before, which makes the algebra of Weyl symbols isomorphic to the operators algebra, is given by

$$(A * B)(\varphi) = A(\varphi) \exp\left[\frac{i\hbar}{2} \overleftarrow{\partial}_a \omega^{ab} \overrightarrow{\partial}_b\right] B(\varphi)$$

$$= \exp\left[\frac{i\hbar}{2} \omega^{ab} \overset{1}{\partial}_a \overset{2}{\partial}_b\right] A(\varphi_1) B(\varphi_2)\big|_{\varphi_1=\varphi_2=\varphi} \qquad (4.5)$$

or more explicitly

$$(A * B)\varphi = \sum_{m=0}^{\infty} \frac{1}{m!} \left(\frac{i\hbar}{2}\right)^m \omega^{a_1 b_1} \cdots \omega^{a_m b_m} (\partial_{a_1} \cdots \partial_{a_m} A)(\partial_{b_1} \cdots \partial_{b_m} B)$$

$$= A(\varphi) B(\varphi) + O(\hbar).$$

So we notice that to the zeroth order in $\hbar$ the star product of two functions reduces to the ordinary pointwise product. For non-zero values of $\hbar$ this multiplication is "*deformed*" in such a way that the resulting star product remains associative but becomes non-commutative in general. Another important ingredient of this formalism is the one called *Moyal brackets* (*mb*) associated to two symbols $A, B \in \mathcal{F}(\mathcal{M}_{2N})$. The (*mb*) are defined as the commutator of the symbols A, B done with respect to the star product:

$$\{A, B\}_{mb} \equiv \frac{1}{i\hbar}(A * B - B * A) = symb\left(\frac{1}{i\hbar}\left[\hat{A}, \hat{B}\right]\right). \qquad (4.6)$$

Using the expression of the star product given in Eqn.(4.5) we get:

$$\{A, B\}_{mb} = A(\varphi) \frac{2}{\hbar} \sin\left[\left(\frac{\hbar}{2}\right) \overleftarrow{\partial}_a \omega^{ab} \overrightarrow{\partial}_b\right] B(\varphi)$$

$$= \{A, B\}_{pb} + O(\hbar^2) + \cdots.$$

So in the classical limit ($\hbar \to 0$) the Moyal brackets reduce to the classical Poisson brackets (*pb*). In mathematical language it is considered a "*deformation*" of the Poisson brackets, which preserves two important properties:

(i) the *mb* obey the Jacobi identity like the *pb*;
(ii) they obeys the Leibniz rule like the *pb*:

$$\{A, B_1 * B_2\}_{mb} = \{A, B_1\}_{mb} * B_2 + B_1 * \{A, B_2\}_{mb}.$$

Note that this correspondence between commutators and Moyal brackets is an *exact correspondence* and it has nothing to do with Dirac Correspondence Principle which says that the Poisson brackets should be replaced in

QM by the commutators. Here we do not have to replace anything because the mb are already the commutators but written in a different language.

Let us now turn to the evolution equation. At the operatorial level we have either the Schrödinger equation for the state or the Heisenberg-von Neumann equation for the density matrix

$$i\,\hbar\partial_t\hat\rho = -\left[\hat\rho, \hat H\right]$$

which, via the *symbol map*, goes into the following equation

$$i\,\hbar\partial_t\rho(\varphi,t) = -\{\rho, H\}_{mb} \tag{4.7}$$

where ρ and H are the symbols of respectively $\hat\rho$ and $\hat H$. For pure states this is the equation of motion for the Wigner functions. In the classical limit $\hbar \to 0$ this equation becomes

$$\partial_t\rho(\varphi,t) = -\{\rho, H\}_{pb} \tag{4.8}$$

which can also be written in the following operatorial form

$$i\partial_t\rho(\varphi,t) = \hat L\,\rho \tag{4.9}$$

with

$$\hat L = i\frac{\partial H}{\partial q}\frac{\partial}{\partial p} - i\frac{\partial H}{\partial p}\frac{\partial}{\partial q} = -i\omega^{ab}\frac{\partial H}{\partial\varphi^b}\frac{\partial}{\partial\varphi^a}$$

where $\hat L$ is known as the *Liouville operator*.

4.1 Marinov's path integral for Wigner functions

Analogously to what has been done in Eqn. (4.8), we can perform the same for Eqn. (4.7)

$$\partial_t\rho = -\{\rho, H\}_{mb}$$
$$\Downarrow$$
$$i\partial_t\rho = \widehat L_\hbar\rho \tag{4.10}$$

where

$$\widehat L_\hbar = -i\sum_{n=0}^{\infty}\frac{2}{\hbar}\frac{(-1)^n}{(2n+1)!}\left[\frac{\hbar}{2}\omega^{ab}\partial_b H\partial_a\right]^{2n+1} \tag{4.11}$$

and this is often called *quantum Liouville operator*. Expanding in $\hbar$ we get

$$\widehat L_\hbar = \widehat L + i\frac{1}{24}\hbar^2\left(\omega^{ab}\partial_b H\right)\left(\omega^{il}\partial_l H\right)\left(\omega^{jm}\partial_m H\right)\partial_a\partial_i\partial_j + O\left(\hbar^3\right) + \cdots.$$

Note that

$$\lim_{\hbar \to 0} \widehat{L}_\hbar = \widehat{L}.$$

We can consider relations (4.9) and (4.10) as the operatorial formulation of classical and quantum mechanics respectively at least for the time evolution. We also know that any operatorial formulation can be turned into a path integral one even for classical systems [Gozzi *et al.* (1989)]. Let us first show how to do this for classical mechanics. Actually this will be done in full details [Gozzi *et al.* (1989)] in Chapter 5, while here we will present a more heuristic derivation. Equation (4.8)

$$i\partial_t \rho = \widehat{L}\rho \tag{4.12}$$

has a formal structure very similar to the Schrödinger equation (with $\hbar = 1$)

$$i\partial_t \psi = \widehat{H}\psi. \tag{4.13}$$

The evolution operator associated to Eqn. (4.13), if we suppose $\widehat{H}$ not explicitly dependent on t, has the form

$$\widehat{U}_q = e^{-i\widehat{H}t}. \tag{4.14}$$

The subindex q is for "*quantum*". Analogously the evolution operator associated to the classical Eqn. (4.12) is

$$\widehat{U}_c = e^{-i\widehat{L}t}, \tag{4.15}$$

where c is for *classical*. We have seen in the first Chaper that the path integral associated to $\widehat{U}_q$ has the form

$$\widehat{U}_q \to \int \mathcal{D}q\mathcal{D}p \, e^{i\int \mathcal{L}\,\mathrm{d}t} \tag{4.16}$$

where $\mathcal{L}$ is the Lagrangian of the system

$$\mathcal{L} = \frac{p^2}{2m} - V(q) \tag{4.17}$$

with p the momentum, q the position and $V(q)$ the potential. This Lagrangian has an associated Hamiltonian

$$H = \frac{p^2}{2m} + V(q) = p\dot{q} - \mathcal{L} \tag{4.18}$$

which is turned into the operator $\widehat{H}$ of Eqn.(4.14) via the replacement

$$p \to -i\frac{\partial}{\partial q}. \tag{4.19}$$

In this manner

$$H = \frac{p^2}{2m} + V(q) \rightarrow \widehat{H} = -\frac{1}{2m}\frac{\partial^2}{\partial q^2} + V(\widehat{q}) . \qquad (4.20)$$

We can somehow say that the path integrals turn the p into an operator $-i\frac{\partial}{\partial q}$. This is related to the quantization process which turns the Poisson bracket $\{q,p\} = 1$ into $[\widehat{q},\widehat{p}] = i$:

$$\{q,p\} \rightarrow \frac{1}{i}[\widehat{q},\widehat{p}] . \qquad (4.21)$$

Let us now perform a set of similar steps for the operator (4.15). The Liouvillian $\widehat{L}$ plays the role of $\widehat{H}$ and we should find which is the "classical" Hamiltonian that produces $\widehat{L}$ as "quantum" Hamiltonian. In $\widehat{L}$, which is

$$\widehat{L} = -i\omega^{ab}\frac{\partial H}{\partial \varphi^b}\frac{\partial}{\partial \varphi^a}, \qquad (4.22)$$

we have both $\frac{\partial}{\partial q}$ and $\frac{\partial}{\partial p}$ as opposed to $\widehat{H}$ where there is only $\frac{\partial}{\partial q}$. So at the path integral level we must have two variables analogous to the p of the quantum case. Let us call these variables as λ_a. They must be "canonically conjugated" to the φ^a so we must introduce a sort of "extended Poisson brackets" [Gozzi *et al.* (1989)] of the form:

$$\{\varphi^a, \lambda_b\} = \delta^a_b . \qquad (4.23)$$

The path integral procedure will turn the brackets (4.23) into the commutators:

$$\{\varphi^a, \lambda_b\} = \delta^a_b \rightarrow \left[\widehat{\varphi}^a, \widehat{\lambda}_b\right] = i\delta^a_b$$

so that $\widehat{\lambda}^b = -i\frac{\partial}{\partial \varphi^b}$. This is *not the quantization* process: no $\hbar$ appears!

Using the λ_a the classical "Hamiltonian", which we will indicate with $\widetilde{\mathcal{H}}_B$, associated to the operator $\widehat{L}$ is

$$\widetilde{\mathcal{H}}_B \equiv \omega^{ab}\frac{\partial H}{\partial \varphi^b}\lambda_a \qquad (4.24)$$

with

$$\omega^{ab} = \begin{pmatrix} 0 & 1 \\ -1 & 0 \end{pmatrix} . \qquad (4.25)$$

With this symplectic matrix we can write the equations of motions of p and q, which are

$$\dot{q} = \frac{\partial H}{\partial p}$$

$$\dot{p} = -\frac{\partial H}{\partial q},$$

as

$$\dot{\varphi}^a = \omega^{ab}\frac{\partial H}{\partial \varphi^b}. \tag{4.26}$$

Going back now to the $\widetilde{\mathcal{H}}_B$ of Eqn.(4.24) we can derive the associated Lagrangian which will be

$$\widetilde{\mathcal{L}}_B = \lambda_a\dot{\varphi}^a - \lambda_a\omega^{ab}\frac{\partial H}{\partial \varphi^b}. \tag{4.27}$$

So, analogously to the quantum case which is Eqn.(4.16), the classical version for Eqn.(4.15) will be

$$\widehat{U}_c \to \int \mathcal{D}\varphi^a \mathcal{D}\lambda_a e^{i\int \widetilde{\mathcal{L}}_B\, dt}. \tag{4.28}$$

We want to stress that this is *not* the quantum version of the classical evolution equation (4.12); it is only the functional version of an operatorial evolution equation like (4.12). That building the path integral does not mean quantizing got very clear in the early 80's, when people built the path integral associated to *classical* stochastic processes. We will not enter in details here but the interested reader can consult [Gozzi (1993)] which is a review and contains references to all the original papers on the subject. Having built the path integral for the classical evolution operator $\widehat{L}$, we can do the same for the quantum Liouville operator: $\widehat{L}_\hbar$ of Eqn.(4.11)

$$\widehat{L}_\hbar = -i\sum_{n=0}^{\infty}\frac{2}{\hbar}\frac{(-1)^n}{(2n+1)!}\left[\frac{\hbar}{2}\omega^{ab}\partial_b H\partial_a\right]^{2n+1}. \tag{4.29}$$

Like we did in $\widehat{L}$, where we replaced $-i\frac{\partial}{\partial\varphi^a}$ with the variables λ_a, we get the following expression for the quantum analogue of $\widetilde{\mathcal{H}}_B$ of Eqn.(4.24) which we shall call $\widetilde{\mathcal{H}}_B^\hbar$:

$$\widetilde{\mathcal{H}}_B^\hbar = \sum_{n=0}^{\infty}\frac{2}{\hbar}\frac{1}{(2n+1)!}\left[\frac{\hbar}{2}\omega^{ab}\partial_b H\lambda_a\right]^{2n+1}. \tag{4.30}$$

The associated Lagrangian will be

$$\widetilde{\mathcal{L}}_B^\hbar = \lambda_a\dot{\varphi}^a - \widetilde{\mathcal{H}}_B^\hbar.$$

Instead of Eqn.(4.30) we will use the following more compact expression which is easy to prove:

$$\widetilde{\mathcal{H}}_B^\hbar = \frac{1}{2\hbar}\left[H\left(\varphi^a - \hbar\omega^{ab}\lambda_b\right) - H\left(\varphi^a + \hbar\omega^{ab}\lambda_b\right)\right] \tag{4.31}$$

where H is the usual Hamiltonian of the system, the one entering the equation of motion (4.26). The $\widetilde{\mathcal{H}}_B^\hbar$ was already present in the literature [Baker

(1958); Bayen *et al.* (1978a,b); Sharan (1979)] but it was independently rederived in [Gozzi and Reuter (1994a)]. Also the variables $\varphi^a - \hbar\omega^{ab}\lambda_b$ and $\varphi^a + \hbar\omega^{ab}\lambda_b$ had already made their appearance in the literature [Bopp (1961); Kubo (1964)] and are known as Bopp variables. A nice review of the whole Moyal formalism can be found in [Marinov (1991)].

Let us now turn to our Lagrangian $\widetilde{\mathcal{L}}_B^\hbar$ which via Eqn.(4.31) can be written as

$$\widetilde{\mathcal{L}}_B^\hbar = \dot{\varphi}^a \lambda_a - \frac{1}{2\hbar}\left[H\left(\varphi^a - \hbar\omega^{ab}\lambda_b\right) - H\left(\varphi^a + \hbar\omega^{ab}\lambda_b\right)\right]. \quad (4.32)$$

This is the same Lagrangian obtained by Marinov in [Marinov (1991)]. He made a different derivation from the one presented here. His derivation applies only to Wigner functions and not, like this one, to any ρ. Our derivation is based on materials contained in [Gozzi *et al.* (1989)] and [Gozzi and Reuter (1994a)]. For the interested reader Marinov used different variables called ξ^a which are related to our variables via

$$\xi^a = \omega^{ab}\lambda_b. \quad (4.33)$$

The variables λ_a have a well known meaning which had escaped Marinov. They are the so-called *response field* variables which were reviewed in [Gozzi (1993)] and used to derive the fluctuation dissipation theorem in a modern way in [Gozzi (1984, 1985)].

4.2 Semiclassical expansion in the Marinov's path integral

If we use the Lagrangian $\widetilde{\mathcal{L}}_B^\hbar$ written in Eqn.(4.32) of the previous section we can build something as the functional generator for the correlation functions and it will read as follows:

$$Z_{QM}^W = \int \mathcal{D}\varphi^a \mathcal{D}\lambda_a e^{i\int \widetilde{\mathcal{L}}_B^\hbar \, dt + i\int J_a \varphi^a \, dt + i\int \widetilde{J}^a \lambda_a \, dt} \quad (4.34)$$

where J_a and $\widetilde{J}^a$ are the currents associated respectively to φ^a and λ_a. The index W on Z_{QM}^W stands for "Wigner" function. Putting the currents to zero we have

$$Z_{QM}^W[0] = \int \mathcal{D}\varphi^a \mathcal{D}\lambda_a e^{i\int \widetilde{\mathcal{L}}_B^\hbar \, dt}$$

where, as in Eqn.(4.32),

$$\widetilde{\mathcal{L}}_B^\hbar = \dot{\varphi}^a \lambda_a - \frac{1}{2\hbar}\left[H\left(\varphi^a - \hbar\omega^{ab}\lambda_b\right) - H\left(\varphi^a + \hbar\omega^{ab}\lambda_b\right)\right]. \quad (4.35)$$

Let us now derive the "equations of motion" by doing the variation of the action associated to Eqn.(4.35). The variation with respect to φ^a gives

$$\partial_t \frac{\partial \widetilde{\mathcal{L}}_B^\hbar}{\partial \dot{\varphi}^a} - \frac{\partial \widetilde{\mathcal{L}}_B^\hbar}{\partial \varphi^a} = 0$$

$$\Downarrow$$

$$\dot{\lambda}_a + \frac{1}{2\hbar} \left[\frac{\partial H \left(\varphi^a - \hbar \omega^{ab} \lambda_b \right)}{\partial \varphi^a} - \frac{\partial H \left(\varphi^a + \hbar \omega^{ab} \lambda_b \right)}{\partial \varphi^a} \right] = 0 \quad (4.36)$$

which is the equation of motion of λ_a. The variation of $\widetilde{\mathcal{L}}_B^\hbar$ with respect to λ_a gives

$$\partial_t \frac{\partial \widetilde{\mathcal{L}}_B^\hbar}{\partial \dot{\lambda}_a} = \frac{\partial \widetilde{\mathcal{L}}_B^\hbar}{\partial \lambda_a}$$

$$\Downarrow$$

$$\dot{\varphi}^a = \frac{1}{2\hbar} \left[\frac{\partial H \left(\varphi^a - \hbar \omega^{ab} \lambda_b \right)}{\partial \lambda_a} - \frac{\partial H \left(\varphi^a + \hbar \omega^{ab} \lambda_b \right)}{\partial \lambda_a} \right] . \quad (4.37)$$

In this last equation the differentation with respect to λ_a can be turned into a differentiation with respect to φ^a by noticing that the arguments in H are $\varphi^a - \hbar \omega^{ab} \lambda_b$ and $\varphi^a + \hbar \omega^{ab} \lambda_b$. In this manner with simple manipulations Eqn. (4.37) can be turned into the following one:

$$\dot{\varphi}^a = \frac{\omega^{ab}}{2} \left[\frac{\partial H \left(\varphi^a - \hbar \omega^{ab} \lambda_b \right)}{\partial \varphi^b} + \frac{\partial H \left(\varphi^a + \hbar \omega^{ab} \lambda_b \right)}{\partial \varphi^b} \right] . \quad (4.38)$$

Note that if we take the limit $\hbar \to 0$ of Eqn. (4.38) we get

$$\dot{\varphi}^a = \omega^{ab} \frac{\partial H \left(\varphi \right)}{\partial \varphi^b} \quad (4.39)$$

which is the standard Hamilton classical equation of motion.

Also Eqn.(4.36) can be put in a different form by multiplying it by $\hbar \omega^{ba}$. The result is

$$\hbar \omega^{ba} \dot{\lambda}_a = -\frac{1}{2} \omega^{ba} \left[\frac{\partial H \left(\varphi^a - \hbar \omega^{ab} \lambda_b \right)}{\partial \varphi^a} - \frac{\partial H \left(\varphi^a + \hbar \omega^{ab} \lambda_b \right)}{\partial \varphi^a} \right] \quad (4.40)$$

and calling ξ^b the quantity $\xi^b = \hbar \omega^{ba} \dot{\lambda}_a$ we get that Eqn. (4.40) becomes

$$\dot{\xi}^a = -\frac{1}{2} \omega^{ab} \left[\frac{\partial H \left(\varphi - \xi \right)}{\partial \varphi^b} - \frac{\partial H \left(\varphi + \xi \right)}{\partial \varphi^b} \right] \quad (4.41)$$

while Eqn.(4.38) is turned into:

$$\dot{\varphi}^a = \frac{\omega^{ab}}{2} \left[\frac{\partial H \left(\varphi - \xi \right)}{\partial \varphi^b} + \frac{\partial H \left(\varphi + \xi \right)}{\partial \varphi^b} \right] . \quad (4.42)$$

Equations (4.41) and (4.42) were already presented in the paper of Marinov [Marinov (1991)].

There is another reason for introducing the variables ξ^a. In fact, in terms of φ^a and ξ^a, the $\widetilde{\mathcal{L}}_B^\hbar$ in Eqn.(4.35) can be written as

$$\widetilde{\mathcal{L}}_B^\hbar = \frac{1}{\hbar} \left\{ \dot{\varphi}^a \omega_{ab} \xi^b - \frac{1}{2} \left[H(\varphi - \xi) - H(\varphi + \xi) \right] \right\}$$

and in this manner the generating functional Z_{QM}^W becomes

$$Z_{QM}^W = \int \mathcal{D}\xi^a \mathcal{D}\varphi^a e^{\frac{i}{\hbar} \int \widetilde{\widetilde{\mathcal{L}}} \, dt} \tag{4.43}$$

where

$$\widetilde{\widetilde{\mathcal{L}}} = \dot{\varphi}^a \omega_{ab} \xi^b - \frac{1}{2} \left[H(\varphi - \xi) - H(\varphi + \xi) \right] . \tag{4.44}$$

Thus $\widetilde{\widetilde{\mathcal{L}}}$ is a Lagrangian which does not contain $\hbar$ anymore and the $\hbar$ in Z_{QM}^W of Eqn. (4.43) is only in front of the the integral. So the limit $\hbar \to 0$ forces the path integral to sit on the equation of motion of $\widetilde{\widetilde{\mathcal{L}}}$ which are the Eqns.(4.41) and (4.42) and that we will rewrite here

$$
\begin{cases}
\dot{\xi}^a = -\dfrac{1}{2} \omega^{ab} \left[\dfrac{\partial H(\varphi - \xi)}{\partial \varphi^b} - \dfrac{\partial H(\varphi + \xi)}{\partial \varphi^b} \right] \\[4mm]
\dot{\varphi}^a = \dfrac{1}{2} \omega^{ab} \left[\dfrac{\partial H(\varphi - \xi)}{\partial \varphi^b} + \dfrac{\partial H(\varphi + \xi)}{\partial \varphi^b} \right]
\end{cases} . \tag{4.45}
$$

Note that *these are not the classical equations of motion*. They become so only if $\xi \to 0$. So ξ plays the role of a "quantum fluctuation". Equations (4.45) are like effective equations which already include some "*quantum effects*".

What Marinov observed is that there are many more solutions of Eqns.(4.45) than the classical ones that are just those for which $\xi^a = 0$ and $\varphi^a = \varphi_{cl}^a$. That this is a solution is clear from the first of Eqns.(4.45) and it is due to the crucial minus sign in the r.h.s. of the first equation. For $\xi = 0$ we obtain the equation $\dot{\xi} = 0$ which has the consistent solution $\xi(t) = 0$.

As there are other solutions of Eqns.(4.45) besides the classical ones, one could think that they may turn out to be useful in order to better understand some quantum phenomenon like for example tunneling. After all Eqns.(4.45) contain already some "*quantum*" effects because they are sort

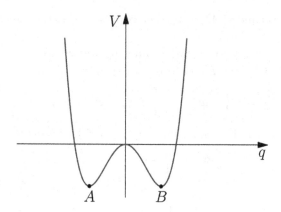

Fig. 4.1   Double well potential.

of "effective" equations as we explained before. With the usual equations of motion, in order to get at the leading order in $\hbar$ for the tunneling effect, people had to introduce *instantons* [Coleman (1985)]. For example if we had a potential like

$$V = -q^2 + q^4 \tag{4.46}$$

which has no classical solution which goes from A to B, there is no manner to get the correction even at the first order in $\hbar$ via the WKB of the standard path integral described in Section 3.3.4. Nevertheless we know, from the operatorial quantum mechanical version of the problem, that for a particle in A there is a non-zero probability of tunneling to B. In the standard path integral formulation the trick is to go to the imaginary time and this turns the potential upside down. This new potential has new "classical" solutions which go from A to B and are indicated in Fig. 4.2 by a dotted line. These "pseudo-classical" solutions are called instantons and are those which give the first quantum corrections in a WKB approximation. To see more precisely how the instantons contribute to the transition probability let us note the following facts. The instantons are localized in the euclidean time which means that the solution stays close to A for a long time and then suddenly jumps close to B and it remains there. Besides the solutions going from A to B, we have to consider the solutions going from B to A, called anti-instantons. The crucial observation is that one can obtain other approximate solutions to the euclidean equations of motion by suitably combining many instanton and anti-instanton solutions with each other. These approximate solutions also contribute to the final amplitude and have

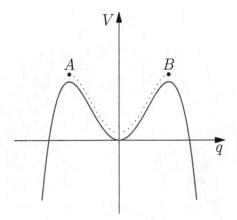

Fig. 4.2 Instanton (dotted line) going from A to B.

to be included. These are the main ideas of the computation at the level of the standard path-integral and they are reported in details in [Coleman (1985)].

We thought that maybe in the Marinov path integral we do not have to do such *"unnatural"* operations such as turning the potential upside down because we have already alternative solutions. The new potential is in two variables q and ξ_q, defined below, and could have the form shown in Fig. 4.3. There could be solutions which go from A to B by staying in the valley of the minima of the potential as indicated in Fig. 4.3 by the dashed line. The mechanics is different from the classical one and it allows trajectories to exist between A and B. This different mechanics can be better understood if we analyze the equation of motion of q and ξ_q. Let us start with the potential (4.46) and let us derive the equation of motion (4.45) for p and q which are the components of φ^a and for ξ_q and ξ_p which are the components of ξ_a. The equations turn out to be

$$\dot{p} = 2q - 2\left(q - \xi_q\right)^2 - 2\left(q + \xi_q\right)^3 \qquad (4.47a)$$

$$\dot{q} = \frac{p}{m} \qquad (4.47b)$$

$$\dot{\xi}_q = \frac{\xi_p}{m} \qquad (4.47c)$$

$$\dot{\xi}_p = 2\xi_p + 2\left(q - \xi_q\right)^3 - 2\left(q + \xi_q\right)^3 . \qquad (4.47d)$$

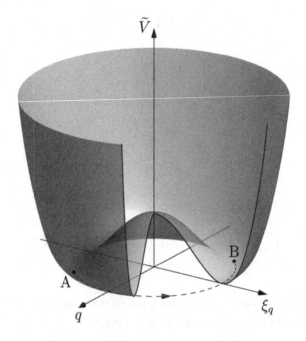

Fig. 4.3

We can eliminate p and ξ_p and go to second order equations which are

$$m\ddot{q} = 2q\left(1 - 6\xi_q^2\right) - 4q^3 \tag{4.48a}$$

$$m\ddot{\xi}_q = 2\xi_q\left(1 - 6\xi_q^2\right) - 4\xi_q^3 . \tag{4.48b}$$

Let us compare the first equation above with the standard classical equation of motion that we would obtain from the potential (4.46):

$$m\ddot{q} = 2q - 4q^3 . \tag{4.49}$$

The difference is in the coefficient of the linear term on the r.h.s. of Eqn.(4.49) and the analogue in the first equation in Eqn.(4.48). In the case of Eqn.(4.49) this coefficient is a fixed quantity and it does not change with time, while in the case of Eqns.(4.48) it is $2\left(1 - 6\xi_q^2\right)$ and it changes with time because ξ_q flows in time. It is as if in Eqn.(4.49) we had a potential

$$V(q) = -q^2 + q^4$$

while in the case of Eqn.(4.48) the potential is

$$\widetilde{V}(q) = -\left(1 - 6\xi_q^2\right)q^2 + q^4 .$$

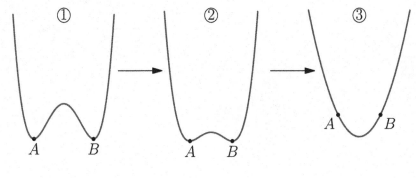

Fig. 4.4

ξ_q was zero at time $t = 0$ so there we would have $\tilde{V}(q) = V(q)$, while time progresses ξ_q would change and we could have a change of the potential like in Fig. 4.4.

So a particle starting in A would at first not be able to overcome the potential barrier in ① of Fig. 4.4 but as soon as time progresses the central peak decreases and the particle find itself with a lower barrier to overcome (② of Fig. 4.4), and as time progresses further it may find itself with no barrier at all like in ③ of Fig. 4.4.

The first trick is to choose the proper solution for ξ_q so that the potential behaves like in Fig. 4.4. Using these solutions we do not have to use instantons anymore. Some further unpublished work [Gozzi and Reuter (1995)] was done on this using both numerical and analytical methods for the infinite well potential with a step in the middle like in Fig. 4.5. Further work is needed in order to complete this project.

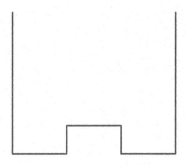

Fig. 4.5

Chapter 5

Classical Mechanics and its associated Path Integral

5.1 The work of Koopman-von Neumann (KvN) on the operatorial version of classical mechanics

We have seen in the previous chapter that classical statistical mechanics has an operatorial version given by the Liouville equation

$$i\frac{\partial}{\partial t}\rho(p,q) = \widehat{L}\rho \tag{5.1}$$

where $\rho(p,q)$ is the classical probability density and $\widehat{L}$ the Liouville operator defined as

$$\widehat{L} = i\left(\frac{\partial H}{\partial q}\frac{\partial}{\partial p} - \frac{\partial H}{\partial p}\frac{\partial}{\partial q}\right). \tag{5.2}$$

$\widehat{L}$ is a differential operator and, as opposed from the Schrödinger one

$$\widehat{H} = -\frac{1}{2}\frac{\partial^2}{\partial q^2} + V(q), \tag{5.3}$$

is first order in the derivative. $\rho(p,q)$, being a probability density, has the property that:

$$\int \rho(p,q)\,\mathrm{d}p\,\mathrm{d}q = 1. \tag{5.4}$$

So, as a function, ρ is an element of $\mathcal{L}^{(1)}$ (integrable functions) and not, like the Schrödinger wave function $\psi(q)$, an element of $\mathcal{L}^{(2)}$ (square integrable functions).

What Koopman and von Neumann [Koopman (1931); von Neumann (1932a,b)] did was somehow to "derive" the Liouville formalism from more basic concepts. They introduced a set of postulates that are the following:

(1) A state of a classical system, whose phase space is indicated by
$\mathcal{M}$ with coordinates p and q, is represented by an element $|\psi\rangle$ of a
Hilbert space $\mathcal{H}$.

(2) On this Hilbert space the operators $\hat{p}$ and $\hat{q}$, whose eigenvalues are
p and q, commute

$$[\hat{p}, \hat{q}] = 0.$$

(3) The states $\langle p, q | \psi \rangle$ are square integrable and their modulus square
$|\psi(p,q)|^2$ is the probability density $\rho(p,q)$ to find the system in
(p,q):

$$|\psi|^2 = \rho.$$

(4) The evolution of $\psi(p,q)$ is given by the Liouville equation

$$i\frac{\partial \psi}{\partial t} = \widehat{L}\psi. \tag{5.5}$$

(5) The observables of the theory are given by Hermitian operators
$O(\hat{p}, \hat{q})$, functions of $\hat{p}$ and $\hat{q}$.

As a consequence of postulates (3) and (4) we get the Liouville evolution
of ρ (the Liouville equation). In fact

$$
\begin{aligned}
\frac{\partial \rho}{\partial t} &= \frac{\partial \psi^*}{\partial t}\psi + \psi^*\frac{\partial \psi}{\partial t} \\
&= \left(-i\widehat{L}\psi^*\right)\psi + \psi^*\left(-i\widehat{L}\psi\right) \\
&= -i\widehat{L}\left(\psi^*\psi\right) = -i\widehat{L}\rho
\end{aligned}
$$

where we have used the fact that $\widehat{L}$ is first order in the derivatives. This
fact does not happen for the Schrödinger operator which is second order in
the derivatives. In this case ψ and ρ do not have the same evolution. In
fact while the *quantum* wave function satisfies the Schrödinger equation

$$i\hbar\frac{\partial \psi}{\partial t} = \widehat{H}\psi \tag{5.6}$$

with

$$\widehat{H} = -\frac{\hbar^2}{2}\frac{\partial^2}{\partial q^2} + V(q),$$

the *quantum* $\rho = |\psi|^2$ satisfies a totally different equation which is

$$\frac{\partial \rho}{\partial t} = -\mathrm{div} J$$

where

$$J = -\frac{i\hbar}{2m} \left(\psi^* \nabla \psi - \psi \nabla \psi^* \right).$$

Another point worth noting is the 5th postulate. This triggers a superselection principle which forbids the *superposition* in CM. This was something needed in a Hilbert space formulation of CM like the KvN is. It was needed because there is superposition in QM but *not* in CM. It has been proven in [Gozzi and Pagani (2010)] that actually the postulate (5) of KvN is not needed. In fact in the enlarged set of variables that we will introduce in the next section, it is enough to require the invariance under a universal gauge symmetry present in the enlarged space. More details will be provided in Section 5.5 or can be found in [Gozzi and Pagani (2010)]. The *non-superposition* in CM was a crucial issue to study and implement in the KvN formalism. We feel that KvN did not address it in a definitive manner and that is why we came back to it in ref. [Gozzi and Pagani (2010)].

5.2 Path Integrals for classical mechanics (CPI) from the KvN formalism

It is well known from stochastic processes [Gozzi (1993)] that if we have an *operatorial* formalism for a theory then there is always an associated *path integral* one. This must also be so for CM because here we have an operatorial formalism which is the KvN one that we have just presented.

Let us just set the formalism with the Hamilton equation of motion written in the form

$$\dot{\varphi}^a = \omega^{ab} \frac{\partial H}{\partial \varphi^b} \tag{5.7}$$

with $a = 1, \cdots, 2n$ and where

$$\varphi^a = \left(q^1, \cdots, q^n, p^1, \cdots, p^n \right).$$

$H(\varphi)$ is the Hamiltonian of the system and ω^{ab} the symplectic matrix

$$\omega^{ab} = \begin{pmatrix} 0 & \mathbb{I} \\ -\mathbb{I} & 0 \end{pmatrix}.$$

The classical distribution $\rho(\varphi)$, like the KvN waves $\psi(\varphi)$, propagates via a kernel which we denote by $K(\varphi_f, t_f | \varphi_i, t_i)$, where φ_f is the final phase space point and φ_i is the initial one. This kernel acts as

$$\psi(\varphi_f, t_f) = \int K(\varphi_f, t_f | \varphi_i, t_i) \, \psi(\varphi_i, t_i) \, \mathrm{d}^{2n}\varphi_i. \tag{5.8}$$

It is easy to realize that this propagator has the form

$$K\left(\varphi_f, t_f | \varphi_i, t_i\right) = \delta\left[\varphi_f - \Phi_{cl}\left(t_f; \varphi_i, t_i\right)\right] \tag{5.9}$$

where $\Phi_{cl}\left(t_f; \varphi_i, t_i\right)$ is the solution of the classical equations of motion (5.7) at time t_f if the system started in φ_i at time t_i. Let us now discretize the time interval $t_f - t_i$ into N intervals and let us rewrite Eqn.(5.9) as follows

$$K\left(\varphi_f, t_f | \varphi_i, t_i\right) = \lim_{N \to \infty} \left\{ \int \prod_{i=1}^{N-1} d\varphi_j \delta\left[\varphi_j\left(t_j\right) - \Phi_{cl}\left(t_j; \varphi_i, t_i\right)\right] \right\}$$
$$\times \delta\left[\varphi_f - \Phi_{cl}\left(t_f; \varphi_i, t_i\right)\right]. \tag{5.10}$$

We have basically inserted a set of identity integrating over points φ_j which are intermediate between the initial and final points φ_i and φ_f. All the Dirac deltas which appear in Eqn.(5.10) can be rewritten in the following manner:

$$\delta\left[\varphi_j\left(t_j\right) - \Phi_{cl}\left(t_j; \varphi_i, t_i\right)\right] = \delta\left[\dot{\varphi}^a - \omega^{ab}\frac{\partial H}{\partial \varphi^b}\right]\Bigg|_{t_j}$$
$$\times \det\left[\delta_b^a \partial_t - \omega^{ac}\frac{\partial^2 H}{\partial \varphi^c \partial \varphi^b}\right]\Bigg|_{t_j} \tag{5.11}$$

where the $\left[\dot{\varphi}^a - \omega^{ab}\frac{\partial H}{\partial \varphi^b}\right]$ is the function whose zeros are the Φ_{cl} of the l.h.s. of Eqn.(5.11). Let us now introduce an auxiliary field λ_a and rewrite the Dirac deltas present on the r.h.s. of Eqn.(5.11) as follows:

$$\delta\left[\dot{\varphi}^a - \omega^{ab}\frac{\partial H}{\partial \varphi^b}\right] = \int d\lambda_a e^{i\lambda_a\left[\dot{\varphi}^a - \omega^{ab}\frac{\partial H}{\partial \varphi^b}\right]}. \tag{5.12}$$

Let us also introduce a set of $4n$ Grassmannian variables c^a, $\bar{c}_a$ (with $a = 1, \cdots, 2n$) and rewrite the determinant present on the r.h.s. of Eqn.(5.11) as follows:

$$\det\left[\delta_b^a \partial_t - \omega^{ac}\frac{\partial^2 H}{\partial \varphi^c \partial \varphi^b}\right] = \int dc^a d\bar{c}_a e^{-\bar{c}_a\left[\delta_b^a \partial_t - \omega^{ac}\frac{\partial^2 H}{\partial \varphi^c \partial \varphi^b}\right]c^b}. \tag{5.13}$$

The reader not expert in Grassmannian variables should first read Appendix D. If we now insert Eqns.(5.12) and (5.13) into Eqn.(5.10) and take the continuum limit in the time discretization we get:

$$K\left(\varphi_f, t_f | \varphi_i, t_i\right) = \int_{\varphi_i}^{\varphi_f} \mathcal{D}''\varphi \mathcal{D}\lambda \mathcal{D}c \mathcal{D}\bar{c}\, e^{i\int dt \widetilde{\mathcal{L}}} \tag{5.14}$$

with $\widetilde{\mathcal{L}}$ given by

$$\widetilde{\mathcal{L}} = \lambda_a\left[\dot{\varphi}^a - \omega^{ab}\frac{\partial H}{\partial \varphi^b}\right] + i\bar{c}_a\left[\delta_b^a \partial_t - \omega^{ac}\frac{\partial^2 H}{\partial \varphi^c \partial \varphi^b}\right]c^b. \tag{5.15}$$

$\mathcal{D}''\varphi$ in Eqn.(5.14) indicates that the functional integration is taken over all trajectories in φ which start at fixed φ_i and end at fixed φ_f while $\mathcal{D}\lambda$, $\mathcal{D}c$ and $\mathcal{D}\bar{c}$ integrate over all the trajectories in their respective space without having fixed the extremities.

Formula (5.14) is what we call the *"path integral for classical mechanics"* [Gozzi *et al.* (1989)] and we shall indicate it with the acronym CPI for *"classical path integral"*. It is of course different from the path integral for quantum mechanics (which we shall indicate with the acronym QPI for *"quantum path integral"*). This last one has a different weight, $\exp \frac{i}{\hbar} \int \mathcal{L} dt$, and a different measure, $\mathcal{D}''q\mathcal{D}p$, compared to the CPI:

$$K_{CPI} = \int \mathcal{D}''\varphi \, \mathcal{D}\lambda \, \mathcal{D}c \, \mathcal{D}\bar{c} \, e^{i \int \tilde{\mathcal{L}} dt} \tag{5.16}$$

$$K_{QPI} = \int \mathcal{D}''q \, \mathcal{D}p \, e^{\frac{i}{\hbar} \int \mathcal{L} dt} \tag{5.17}$$

Note also that $\int \tilde{\mathcal{L}} dt$ does not have the dimension of an action as $\int \mathcal{L} dt$ has in the QPI. $\int \tilde{\mathcal{L}} dt$ is a dimensionless quantity, a pure phase.

It is clear that the "strange" Lagrangian $\tilde{\mathcal{L}}$ forces the trajectories in φ to lie on the classical one like it was originally in Eqn.(5.9). It is also easy to note that the Lagrangian $\tilde{\mathcal{L}}$ in Eqn.(5.17) provides the same equations of motion for φ like the standard Lagrangian $\mathcal{L}$:

$$\mathcal{L} = \dot{q}p - \frac{p^2}{2} - V(q) . \tag{5.18}$$

In fact from $\tilde{\mathcal{L}}$, doing the variation with respect to λ_a, we get the equation of motion

$$\dot{\varphi}^a - \omega^{ab} \frac{\partial H}{\partial \varphi^b} = 0 . \tag{5.19}$$

Different from $\mathcal{L}$, the $\tilde{\mathcal{L}}$ provides the equation of motion also of the other variables. Varying with respect to $\bar{c}$ we get the equation for c

$$\left[\delta_b^a \partial_t - \omega^{ac} \frac{\partial^2 H}{\partial \varphi^c \partial \varphi^b} \right] c^b = 0 . \tag{5.20}$$

Varying with respect to c we get the equation for $\bar{c}$:

$$\left[\delta_b^a \partial_t + \omega^{ac} \frac{\partial^2 H}{\partial \varphi^c \partial \varphi^b} \right] \bar{c}_a = 0 \tag{5.21}$$

and varying with respect to φ we get the equation of motion for λ:

$$\dot{\lambda}_b + \omega^{ac} \partial_c \partial_b H \lambda_a + i\bar{c}_a \omega^{ac} \partial_c \partial_d \partial_b H c^d = 0 . \tag{5.22}$$

In this enlarged $8n$ dimensional space we can easily build a Hamiltonian formalism. It is a "constrained one". In fact the momenta π_{φ^a} conjugate to φ^a, defined as

$$\pi_{\varphi^a} \equiv \frac{\partial \widetilde{\mathcal{L}}}{\partial \dot{\varphi}^a} \tag{5.23}$$

are equal to λ_a:

$$\pi_{\varphi^a} = \lambda_a \tag{5.24}$$

and analogously for the $\bar{c}_a$ with the momenta conjugate to c^a. Anyhow we do not have to invoke the Dirac formalism for constrained systems to derive that the Hamiltonian associated to $\widetilde{\mathcal{L}}$ is

$$\widetilde{\mathcal{H}} = \lambda_a \omega^{ab} \frac{\partial H}{\partial \varphi^b} + i\bar{c}_a \omega^{ac} \frac{\partial^2 H}{\partial \varphi^c \partial \varphi^b} c^b \tag{5.25}$$

and the extended Poisson brackets (epb) in this space are

$$\left\{ \varphi^a, \varphi^b \right\}_{epb} = 0 \tag{5.26}$$

$$\left\{ \varphi^a, \lambda_b \right\}_{epb} = \delta_b^a \tag{5.27}$$

$$\left\{ \bar{c}_b, c^a \right\}_{epb} = -i\delta_b^a . \tag{5.28}$$

The evolution of a quantity $O\left(\varphi, \lambda, c, \bar{c}\right)$ is given by

$$\frac{d}{dt} O = \left\{ O, \widetilde{\mathcal{H}} \right\}_{epb} . \tag{5.29}$$

The reader may wonder at this point if the CPI of Eqn. (5.14) reproduce the KvN operatorial formalism. To check that we have to first derive from Eqn.(5.14) the "operatorial" version of the variables φ, λ, c and $\bar{c}$. To do that let us derive their "commutators". The commutators, following the work of Feynman [Schulman (1981)], are defined via the path integral in the following manner:

$$\langle [O_1(t), O_2(t)] \rangle = \lim_{\varepsilon \to 0} \langle O_1(t-\varepsilon) O_2(t) - O_2(t-\varepsilon) O_1(t) \rangle \tag{5.30}$$

where $\langle \rangle$ indicates the path integral average, the O_1 and O_2 are two functions of the variables φ, λ, c and $\bar{c}$ and ε is an infinitesimal interval of time. If we use the definition (5.30) and Eqn.(5.14) we get the following commutators

$$\left[\varphi^a, \varphi^b \right] = 0 \tag{5.31}$$

$$\left[\varphi^a, \lambda_b \right] = i\delta_b^a \tag{5.32}$$

$$\left[\bar{c}_a, c^b \right] = \delta_b^a \tag{5.33}$$

while all the others are zero. Note that Eqn. (5.31) indicates that q and p commute in the CPI like it should be in a classical system. Note also that Eqns.(5.31), (5.32), (5.33) are different from the extended Poisson brackets of Eqns.(5.25), (5.26) and (5.27). The commutator (5.33) between $\bar{c}$ and c are graded commutators [Dewitt (1992)]. From the commutators above we see that λ_a can be implemented as

$$\lambda_a = -i\frac{\partial}{\partial\varphi^a} \tag{5.34}$$

and

$$\bar{c}_a = \frac{\partial}{\partial c^a}. \tag{5.35}$$

If we neglect for a moment the Grassmannian variables and restrict ourselves to the φ, λ variables in Eqn.(5.14), we get:

$$\int \mathcal{D}''\varphi \mathcal{D}\lambda e^{i\int \tilde{\mathcal{L}}_B},$$

where $\tilde{\mathcal{L}}_B$ is just the $\tilde{\mathcal{L}}$ that we encountered in Chapter 4:

$$\tilde{\mathcal{L}}_B = \lambda_a\dot{\varphi}^a - \tilde{\mathcal{H}}_B. \tag{5.36}$$

Then, analogously to what we do in quantum mechanics, the associated operatorial formalism is

$$\int \mathcal{D}''\varphi \mathcal{D}\lambda e^{i\int \tilde{\mathcal{L}}_B} \to e^{-i\widehat{\tilde{\mathcal{H}}}_B t} \tag{5.37}$$

where $\widehat{\tilde{\mathcal{H}}}_B$ is the operator obtained from $\tilde{\mathcal{H}}_B$ by replacing λ_a with its operatorial realization (5.34):

$$\widehat{\tilde{\mathcal{H}}}_B = -i\frac{\partial}{\partial\varphi^a}\omega^{ab}\frac{\partial H}{\partial\varphi^b} \tag{5.38}$$

$$= -i\frac{\partial H}{\partial p}\frac{\partial}{\partial q} + i\frac{\partial H}{\partial q}\frac{\partial}{\partial p} = \widehat{L}$$

and this is exactly the Liouville operator of equation Eqn.(5.2). So the correspondence (5.37) turns out to be:

$$\int \mathcal{D}''\varphi \mathcal{D}\lambda e^{i\int \tilde{\mathcal{L}}_B} \to e^{-i\widehat{L}t} \tag{5.39}$$

and the r.h.s. is exactly the KvN formalism. This proves that behind the KvN formalism there is just a Dirac delta weight like in Eqn.(5.9), while behind the quantum formalism there is the Feynman weight $e^{\frac{i}{\hbar}\int \mathcal{L}dt}$.

Another issue to notice is that going from the extended Poisson brackets of Eqns.(5.26), (5.27), (5.28) to the commutator (5.31), (5.32), (5.33) cannot be interpreted as a "quantization". In fact it does not produce the Moyal formalism (which is the quantum analogue of the KvN) but only the classical one. So from a classical formalism we go to another classical framework (an operatorial one).

5.3   Cartan calculus via the CPI

The reader may wonder why we need all these $8n$ variables $(\varphi, \lambda, c, \bar{c})$ to do classical mechanics when we know that we can do it with only $2n$ that are the usual phase space coordinates φ^a.

This is quite correct and the $8n$ variables we are using are somehow *redundant*. The signal of this fact is given by some universal symmetries present in the extended space $(\varphi, \lambda, c, \bar{c})$ but not in the restricted one parametrized by (φ). These universal symmetries are given by the following charges:

$$Q_{BRS} = ic^a \lambda_a \tag{5.40a}$$

$$\bar{Q}_{BRS} = i\bar{c}_a \omega^{ab} \lambda_b \tag{5.40b}$$

$$Q_g = c^a \bar{c}_a \tag{5.40c}$$

$$K = \frac{1}{2}\omega_{ab}c^a c^b \tag{5.40d}$$

$$\bar{K} = \frac{1}{2}\omega^{ab}\bar{c}_a \bar{c}_b \tag{5.40e}$$

$$Q_H = ic^a \lambda_a - c^a \partial_a H \tag{5.40f}$$

$$\bar{Q}_H = i\bar{c}_a \omega^{ab} \lambda_b + \bar{c}_a \omega^{ab} \partial_b H . \tag{5.40g}$$

The algebra of these charges, if we introduce also the operator $\widehat{\widetilde{\mathcal{H}}}$ (with its Grassmannian part included) is closed and it is the following:

$$[Q_{BRS}, \bar{Q}_{BRS}] = 0 \tag{5.41a}$$

$$[Q_{BRS}, Q_g] = Q_{BRS} \tag{5.41b}$$

$$[\bar{Q}_{BRS}, Q_g] = -\bar{Q}_{BRS} \tag{5.41c}$$

$$[K, \bar{K}] = Q_g - 1 \tag{5.41d}$$

$$[K, \bar{Q}_{BRS}] = Q_{BRS} \tag{5.41e}$$

$$[K, Q_{BRS}] = 0 \tag{5.41f}$$

$$[Q_H, \bar{Q}_H] = 2i\widehat{\widetilde{\mathcal{H}}} \tag{5.41g}$$

$$[Q_g, K] = 2K \tag{5.41h}$$

$$[Q_g, \bar{K}] = -2K . \tag{5.41i}$$

Note that Eqn.(5.41g) is the same algebra as an $N = 2$ *supersymmetry* and from now on we will call Q_H and $\bar{Q}_H$ supersymmetry generators. It is easy to prove that all the charges can be generated by just four charges via the graded commutators. One of these is just the original Hamiltonian and can be removed. So the left-over independent charges are just three. If the system is made of n non-coupled subsystems then the number is $3n$. This is the right number we need considering that, if at the phase space level we had to go from $8n$ to $2n$ variables, at the configurational level we have to go from $4n$ to n, so we just need $3n$ constants of motion like we have. The problem is to repeat the same procedure for a system with n coupled subsystems. For these we only have the three constants mentioned above and we need to find $3n - 3$ extra conserved charges. Work is in progress on this issue.

If the system has other constants of motion, besides the energy, then we were able [Gozzi and Reuter (1989)] to build further conserved quantities which anyhow do not play any role in the reduction of the number of variables from $8n$ to $2n$. Extra constants of motion can also be built when we have a time-reparametrization invariant formulation of the dynamics [Gozzi (1995); Thacker (1997a)].

It is important to stress that the extra variables λ_a, c^a, $\bar{c}_a$, even if they are redundant, have a physical and geometrical role that we will try to explain in what follows. Let us start with the c^a variables. Let us look at their equation of motion (5.20). They are the same equation of motion as the so-called *Jacobi fields*, or first-variations. Let us choose two nearby classical trajectories $\varphi_{cl}^{(1)}$ and $\varphi_{cl}^{(2)}$ and let us indicate with $\delta\varphi$ the following quantity

$$\delta\varphi(t) = \varphi_{cl}^{(2)}(t) - \varphi_{cl}^{(1)}(t) \, .$$

Doing the first variation of the equation of motion of φ, i.e Eqn.(5.19), we get:

$$\left(\delta_b^a \partial_t - \omega^{ac} \frac{\partial^2 H}{\partial\varphi^c \partial\varphi^b} \right) (\delta\varphi^b) = 0 \, . \tag{5.42}$$

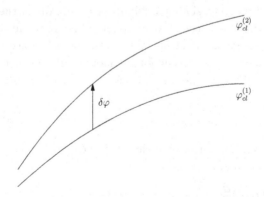

Fig. 5.1   The Jacobi field $\delta\varphi$ describes how two nearby trajectories diverge or approach each other.

Note that this is the same as Eqn.(5.20) which is the equation of motion for the c^a. So $c^a(t)$ can be identified with the Jacobi field. Correlations of these variables gives the Lyapunov exponents and several nice results have been obtained [Gozzi *et al.* (1992a); Gozzi and Reuter (1993, 1994b); Gozzi (1994); Thacker (1997b); Surin and Kurchan (2004)], among which the proof of the Parisi–Ruelle conjecture.

Besides this physical meaning, the c^a have also a geometrical interpretation. Let us do an infinitesimal time translation for $c^a(t)$ using their equation of motion (5.20):

$$c'^a(t + \Delta t) = c^a(t) + \Delta t\, \omega^{ab} \partial_b \partial_d H c^d \qquad (5.43)$$

$$= \frac{\partial \varphi'^a}{\partial \varphi^b} c^b \qquad (5.44)$$

where $\varphi'^a(t) = \varphi^a(t) + \Delta t\, \omega^{ab} \partial_b H$. Note that the c^a in Eqn.(5.44) transforms as differential forms $d\varphi^a$ do. Moreover, like for differential forms, we have to introduce a wedge product [Abraham and Marsden (1978)] between them: $d\varphi^a \wedge d\varphi^b$. We do not have to do that for the c^a because they are already naturally endowed with an anticommuting product $c^a c^b = -c^b c^a$ due to their Grassmannian character.

If we look now at the $\bar{c}_a$ and how they transform under a time translation we get

$$\bar{c}'_a = \frac{\partial \varphi^b}{\partial \varphi'^a} \bar{c}_b \,, \qquad (5.45)$$

which means they transform as the $\frac{\partial}{\partial \varphi^a}$, i.e. as the basis of the antisymmetric tensors.

These features of the c^a and $\bar{c}_a$ generalize to all the symplectic transformations as proven in [Gozzi *et al.* (1992b)]. We will skip for the moment the role of the λ_a which has been studied in detail in [Gozzi and Regini (2000)] together with the geometrical meaning of the overall space coordinatized by the variables $(\varphi, \lambda, c, \bar{c})$. Using for the moment the geometrical meaning of c and $\bar{c}$ we can for example turn an n-form $F_{ab \cdots n}(\varphi)\, d\varphi^a d\varphi^b \cdots d\varphi^n$ into a function of φ and c:

$$F_{ab \cdots n}(\varphi)\, d\varphi^a d\varphi^b \cdots d\varphi^n \to F_{ab \cdots n}(\varphi)\, c^a c^b \cdots c^n. \qquad (5.46)$$

Analogously, given an antisymmetric tensor field $T^{ab \cdots n} \partial_a \partial_b \cdots \partial_n$ we can turn it into a function of φ and $\bar{c}$

$$T^{ab \cdots n}(\varphi)\, \frac{\partial}{\partial \varphi^a} \frac{\partial}{\partial \varphi^b} \cdots \frac{\partial}{\partial \varphi^n} \to T^{ab \cdots n}(\varphi)\, \bar{c}_a \bar{c}_b \cdots \bar{c}_n. \qquad (5.47)$$

Along the same lines all the operation of the Cartan calculus [Abraham and Marsden (1978)] can be turned into operations on our enlarged space implemented by our charges (5.40) and our commutators (5.31), (5.32), (5.33). One of this operation is for example the one of doing the exterior derivative d of an n-form F. The correspondence with our formalism is the following

$$dF \to \left[Q_{BRS}, \widehat{F}\right] \tag{5.48}$$

where $\widehat{F}$ is the function F with the form $d\varphi$ replaced by the Grassmannian variable. Another well known operation is the *interior contraction* [Abraham and Marsden (1978)] of an n-form with a vector field v. This operation is indicated by $i_v F$. The correspondence with our formalism is

$$i_v F \to \left[\widehat{v}, \widehat{F}\right] \tag{5.49}$$

where $\widehat{F}$ was defined before while $\widehat{v}$ is the vector field with $\frac{\partial}{\partial \varphi^a}$ replaced by $\bar{c}_a$. Another operation [Abraham and Marsden (1978)] in symplectic geometry is the one which turn a 1-form, like dH, into a vector field indicated by $(dH)^{\#}$, this is achieved in our formalism in this way

$$(dH)^{\#} \to [\bar{Q}_{BRS}, H] . \tag{5.50}$$

In general the duality present in symplectic geometry between vector fields and forms is achieved in our formalism via the charges K and $\bar{K}$ of Eqns.(5.40d) and (5.40e). Details can be found in [Gozzi *et al.* (1989)].

Another important concept in differential geometry is the one of Lie derivative along a vector field v, which is defined as

$$\mathcal{L}_v \equiv di_v + i_v d . \tag{5.51}$$

If the vector field v is the Hamiltonian vector field $(dH)^{\#}$ then Eqn.(5.51) is written as

$$\mathcal{L}_{(dH)^{\#}} \equiv di_{(dH)^{\#}} + i_{(dH)^{\#}} d . \tag{5.52}$$

In our formalism the action of $\mathcal{L}_{(dH)^{\#}}$ on a form F is given by the correspondence:

$$\mathcal{L}_{(dH)^{\#}} F \to \left[i\widehat{\widetilde{\mathcal{H}}}, \widehat{F}\right]$$

where $\widehat{\widetilde{\mathcal{H}}}$ is the Hamiltonian (5.25). So this explains what is the full Hamiltonian $\widetilde{\mathcal{H}}$ with the Grassmannian variables included: it is the Lie derivative of the Hamiltonian flow, while without the Grassmannian variables it is the

Liouville operator which is also known as the Hamiltonian vector field. So the full $\widetilde{\mathcal{H}}$ dictates the evolution not only of the KvN waves $\psi(\varphi)$ but also of more general objects $\psi(\varphi, c)$ which are differential forms.

Another concept which can be translated into our language is the one of Lie brackets and its generalizations like the Frolicher-Nijenhuis and Schouten-Nijenhuis brackets. All of them become very simple in our formalism and details can be learned from [Gozzi and Mauro (2000)]. Even the supersymmetry charges Q_H and $\bar{Q}_H$ have a geometrical meaning related to the equivariant exterior derivative introduced by Cartan [Abraham and Marsden (1978)] and the supersymmetry commutation relation (5.41g) is known as the Cartan triplet.

All this formalism, like the Lie derivative (5.51), allows to write things in a coordinate-free form. By "coordinate" we mean coordinates of the phase space that are the φ^a. The price we pay in order to have everything written in a coordinate-free way is that we have to enlarge our space.

The formalism of the CPI has been applied also to field theory [Gozzi and Mauro (2002); Gozzi et al. (2005); Carta et al. (2006); Gozzi and Penco (2011); Cattaruzza et al. (2011)]. The reason to do that is that in heavy ion collisions and in the early stage of the universe the high density and high temperature effects killed the quantum ones, so the effective theory is basically a classical one, to which we have to apply both perturbative and non-pertubative approximations and the CPI is the right tool to do that.

5.4   Geometric quantization

The reader may wonder how we can pass from the CPI to the QPI which means how we quantize the system. This is equivalent to passing from the Lie derivative $\mathcal{L}_{(dH)^\#}$ of the Hamiltonian flow (CPI) to the Schrödinger operator (QPI). Note that this is different than going from the classical Hamiltonian to the Schrödinger operator via the Dirac correspondence rules. The transition from the Lie derivative to the Schrödinger operator was done in the 70's and goes under the name of geometric quantization (for a review see [Woodhouse (1997)]). In this chapter we will perform the same kind of quantization but via path integral. Basically we will give a set of rules to pass from the CPI to the QPI and vice versa. We will see that this set of rules are much simpler and more geometrical than those of standard geometric quantization. Moreover the same rules apply both for the states and for the observables while the original geometric quantization had two different sets of rules for the states and the observables [Woodhouse (1997)].

The tools that are needed for implementing our geometric quantization are those of supertime and superfields which we will introduce shortly. First let us turn to the commutation relations we derived from the CPI in Section 5.2 where we used the hat to indicate the operators. The commutation relations were

$$\left[\widehat{\varphi}^a, \widehat{\lambda}_b\right] = i\delta_b^a \tag{5.53a}$$

$$\left[\widehat{c}^a, \widehat{\bar{c}}_b\right] = \delta_b^a \tag{5.53b}$$

and we built a representation of them which is the following: the operator $\widehat{\lambda}_a$ and $\widehat{\bar{c}}_a$ are realized as derivative operators:

$$\widehat{\lambda}_a = -i\frac{\partial}{\partial\varphi^a} \tag{5.54a}$$

$$\widehat{\bar{c}}_a = \frac{\partial}{\partial c^a} \tag{5.54b}$$

while the $\widehat{\varphi}^a$ and $\widehat{c}^a$ are realized as multiplicative operators:

$$\widehat{\varphi}^a|\varphi^a, c^a\rangle = \varphi^a|\varphi^a, c^a\rangle \tag{5.55a}$$

$$\widehat{c}^a|\varphi^a, c^a\rangle = c^a|\varphi^a, c^a\rangle. \tag{5.55b}$$

Looking at the commutation relations (5.53) and indicating the components of $\widehat{\varphi}^a = (\widehat{q}, \widehat{p})$ and of $\widehat{\lambda}_a = \left(\widehat{\lambda}_q, \widehat{\lambda}_p\right)$, we note immediately that we could have made the choice of implementing $\widehat{q}$ and $\widehat{\lambda}_p$ (which commute) as multiplicative operators and the $\widehat{p}$ and $\widehat{\lambda}_q$ as derivative ones:

$$\widehat{p} = i\frac{\partial}{\partial\lambda_p}, \quad \widehat{\lambda}_q = -i\frac{\partial}{\partial q} \tag{5.56}$$

and analogously for the

$$\widehat{c}^a \equiv (\widehat{c}^q, \widehat{c}^p), \quad \widehat{\bar{c}}_a \equiv \left(\widehat{\bar{c}}_q, \widehat{\bar{c}}_p\right).$$

In this case the analog of Eqn.(5.55) would be

$$\widehat{q}|q, \lambda_p, c^q, \bar{c}_p\rangle = q|q, \lambda_p, c^q, \bar{c}_p\rangle \tag{5.57a}$$

$$\widehat{\lambda}_p|q, \lambda_p, c^q, \bar{c}_p\rangle = \lambda_p|q, \lambda_p, c^q, \bar{c}_p\rangle \tag{5.57b}$$

$$\widehat{c}^q|q, \lambda_p, c^q, \bar{c}_p\rangle = c^q|q, \lambda_p, c^q, \bar{c}_p\rangle \tag{5.57c}$$

$$\widehat{\bar{c}}_p|q, \lambda_p, c^q, \bar{c}_p\rangle = \bar{c}_p|q, \lambda_p, c^q, \bar{c}_p\rangle. \tag{5.57d}$$

The two basis (5.55) and (5.57) are related to each other by a Fourier transform and we will come back to this later on.

From now on we follow closely the paper [Abrikosov Jr. *et al.* (2005)].
The reader may dislike the plethora of variables $(\varphi^a, \lambda_a, c^a, \bar{c}_a)$ that make up
the *extended phase space* described above. Actually, thanks to the beautiful
geometry underlying this space, we can assemble together the $8n$ variables
$(\varphi^a, \lambda_a, c^a, \bar{c}_a)$ in a single object as if they were the components of a multiplet. In order to do that we have to first introduce two Grassmann partners
$(\theta, \bar{\theta})$ of the time t. The triplet

$$(t, \theta, \bar{\theta}) \tag{5.58}$$

is known as *supertime* and it is, for the point particle dynamics, the analog
of the *superspace* introduced in supersymmetric field theories [West (1986)].
The object that we mentioned above and which assembles together the $8n$
variables $(\varphi, \lambda, c, \bar{c})$ is defined as

$$\Phi^a(t, \theta, \bar{\theta}) \equiv \varphi^a(t) + \theta c^a(t) + \bar{\theta}\omega^{ab}\bar{c}_b(t) + i\bar{\theta}\theta\omega^{ab}\lambda_b(t). \tag{5.59}$$

We could call the Φ^a *super phase space* variables because their first components φ^a are the standard phase space variables of the system. The
Grassmann variables θ, $\bar{\theta}$ are complex in the sense in which the operation
of complex conjugation can be defined [Dewitt (1992)] for Grassmann variables. Further details can be found in Appendix E, where we also study the
dimensions of these Grassmann variables. The main result is that, even if
there is a lot of freedom in choosing these dimensions, the combination $\theta\bar{\theta}$
has always the dimensions of an action. Using the superfields the full set
of relations (5.26)...(5.28) can be written in a compact form as:

$$\{\Phi^a(t, \theta, \bar{\theta}), \Phi^b(t, \theta', \bar{\theta}')\}_{epb} = -i\omega^{ab}\delta(\bar{\theta} - \bar{\theta}')\delta(\theta - \theta').$$

The space coordinatized by the variables $(\widehat{\varphi}^a, \widehat{c}^a, \widehat{\bar{c}}_a, \widehat{\lambda}_a)$ can be considered as the *"target space"* (in modern language), while the *"base space"* is
given by the supertime $(t, \theta, \bar{\theta})$. It is then natural to ask whether the operators (5.40) are the "representation" on the target space of some differential
operator acting on the base space. The answer is yes and the form of these
differential operators is easy to obtain if we impose on the superfield (5.59)
to be a *scalar* under the symmetry of Eqn.(5.40), i.e.

$$\widehat{\Phi}'^a(t', \theta', \bar{\theta}') = \widehat{\Phi}^a(t, \theta, \bar{\theta}). \tag{5.60}$$

If we indicate with $\widehat{O}$ any of the charges in Eqn.(5.40) and with $\widehat{\mathcal{O}}$ the
corresponding differential operator on the base space, then relation (5.60)
is equivalent to the following one

$$\widehat{\mathcal{O}}\Phi^a = [\widehat{\Phi}^a, \widehat{O}].$$

From this equation it is easy to derive the form of the various operators $\widehat{\mathcal{O}}$. We indicate below the expressions of some of them:

$$\widehat{Q}_{BRS} = -\frac{\partial}{\partial\theta}$$

$$\widehat{\bar{Q}}_{BRS} = \frac{\partial}{\partial\bar{\theta}}$$

$$\widehat{Q}_H = -\frac{\partial}{\partial\theta} - \bar{\theta}\frac{\partial}{\partial t} \qquad (5.61)$$

$$\widehat{\bar{Q}}_H = \frac{\partial}{\partial\bar{\theta}} + \theta\frac{\partial}{\partial t}$$

$$\widehat{\mathcal{H}} = i\frac{\partial}{\partial t}.$$

From these expressions one sees that the BRS and anti-BRS operators behave as translation operators in θ and $\bar{\theta}$. Instead the supersymmetry (SUSY) operators $\widehat{Q}_H$ and $\widehat{\bar{Q}}_H$ turn t into combinations of t with θ and $\bar{\theta}$, which is why we called θ and $\bar{\theta}$ "partners" of t. Once they are applied in sequence, these SUSY operators perform a time translation, as shown by the commutator

$$[\widehat{Q}_H, \widehat{\bar{Q}}_H] = -2\frac{\partial}{\partial t}.$$

The supertime (5.58) is somehow made of three "coordinates" $(t,\theta,\bar{\theta})$. So two different instants in supertime will have respectively coordinates $(t_1,\theta_1,\bar{\theta}_1)$ and $(t_2,\theta_2,\bar{\theta}_2)$. A natural question to ask is the following: is there an interval between those two *super-instants* which is left invariant under the supersymmetry transformations? The answer is yes and actually there is more than one such interval. For example we can take:

$$S \equiv t_2 - t_1 + \theta_2\bar{\theta}_1 - \theta_1\bar{\theta}_2. \qquad (5.62)$$

This expression generalizes the usual interval of time $(t_2 - t_1)$, which is invariant under a global time translation. Also the interval (5.62) is invariant under global time translations just because, as it is invariant under SUSY, it is also invariant if SUSY is performed twice, which is a time translation. All this formalism on time and supertime is familiar to people working on supersymmetry but it may not be so familiar to those working on issues related to quantization. More details can be found in [Abrikosov Jr. *et al.* (2005)].

A further aspect of the Grassmann partners of time $(\theta,\bar{\theta})$ which is worth exploring is the following one. We know that, for an operator theory, we can define either the Heisenberg picture or the Schrödinger one. In the

KvN version of CM the operators, which will be indicated respectively as $\hat{O}_H(t)$ and $\hat{O}_s$ in the two pictures, are related to each other in the following manner:

$$\hat{O}_H(t) \equiv \exp\left[i\widehat{\mathcal{H}}t\right]\hat{O}_s \exp\left[-i\widehat{\mathcal{H}}t\right].$$

One question to ask is what happens if we build the Heisenberg picture generated by the partners of time $\theta, \bar{\theta}$. The analog of the time translation operator for θ and $\bar{\theta}$ is given by $\widehat{Q}_{BRS}$ and $\widehat{\bar{Q}}_{BRS}$ respectively (see Eqn.(5.61)) and so the Heisenberg picture in θ, $\bar{\theta}$ of an operator $\hat{O}_s$ is[1]:

$$\hat{O}_H(\theta,\bar{\theta}) \equiv \exp\left[\theta\widehat{Q} + \widehat{\bar{Q}}\bar{\theta}\right]\hat{O}_s \exp\left[-\theta\widehat{Q} - \widehat{\bar{Q}}\bar{\theta}\right], \qquad (5.63)$$

where we have dropped the "BRS" suffix from $\widehat{Q}$ and $\widehat{\bar{Q}}$. A simple example to start from is the phase space operator $\widehat{\varphi}^a(t)$ which does not depend on $\theta,\bar{\theta}$, so it could be considered as an operator in the Schrödinger picture with respect to $\theta,\bar{\theta}$ and in the Heisenberg picture with respect to t. Its Heisenberg picture version in $\theta,\bar{\theta}$ can be worked out easily (see Appendix F) and the result is:

$$\widehat{\varphi}^a_H(t) \equiv \exp\left[\theta\widehat{Q} + \widehat{\bar{Q}}\bar{\theta}\right]\widehat{\varphi}^a_s(t)\exp\left[-\theta\widehat{Q} - \widehat{\bar{Q}}\bar{\theta}\right] = \widehat{\Phi}^a(t,\theta,\bar{\theta}).$$

This means that the *superphase space operators* $\widehat{\Phi}^a$ *can be considered as the Heisenberg picture* version in θ, $\bar{\theta}$ of the phase space operators. The same holds for any function G of the operators $\widehat{\varphi}$, i.e.:

$$\exp\left[\theta\widehat{Q} + \widehat{\bar{Q}}\bar{\theta}\right]G(\widehat{\varphi}^a)\exp\left[-\theta\widehat{Q} - \widehat{\bar{Q}}\bar{\theta}\right] = G(\widehat{\Phi}^a), \qquad (5.64)$$

see Appendix F for details. In particular, if the function $G(\widehat{\varphi}^a)$ is the Hamiltonian $H(\widehat{\varphi}^a)$ we get from Eqn.(5.64)

$$SH(\widehat{\varphi}^a)S^{-1} = H[\widehat{\Phi}^a] \qquad (5.65)$$

where $S = \exp[\theta\widehat{Q} + \widehat{\bar{Q}}\bar{\theta}]$.

At this point it is instructive to expand the r.h.s. of Eqn.(5.65) in terms of θ, $\bar{\theta}$. We get:

$$H[\widehat{\Phi}^a] = H[\widehat{\varphi}^a] + \theta\widehat{N} + \widehat{\bar{N}}\bar{\theta} - i\bar{\theta}\theta\widehat{\bar{\mathcal{H}}} \qquad (5.66)$$

where

$$\widehat{N} = \widehat{c}^a\partial_a H(\widehat{\varphi}), \qquad \widehat{\bar{N}} = \widehat{\bar{c}}_a\omega^{ab}\partial_b H(\widehat{\varphi})$$

---

[1] With the scalar product under which the Grassmann operators $\widehat{c}$ and $\widehat{\bar{c}}$ are Hermitian [Deotto *et al.* (2003a,b)], the operators $\exp\widehat{\bar{Q}}\bar{\theta}$ and $\exp\theta\widehat{Q}$ are unitary, see Appendix E for further details.

are further conserved charges [Gozzi *et al.* (1989)] (they are linear combinations of those in Eqn.(5.40)). The expansion (5.66) holds also if we replace the operators with the corresponding *c*-number variables, i.e.:

$$H[\Phi^a] = H[\varphi^a] + \theta N + \bar{N}\bar{\theta} - i\bar{\theta}\theta\,\widetilde{\mathcal{H}}(\varphi^a, \lambda_a, c^a, \bar{c}_a). \qquad (5.67)$$

It is interesting to note that the first term in the expansion in $\theta, \bar{\theta}$ on the r.h.s. of Eqn.(5.67) is $H(\varphi^a)$ which generates the dynamics in the standard *phase space* φ^a while the last term is $\widetilde{\mathcal{H}}$, which generates the dynamics in the extended phase space $(\varphi^a, \lambda_a, c^a, \bar{c}_a)$.

Another interesting point to study is the equations of motion in the Heisenberg picture with respect to θ, $\bar{\theta}$. Using the commutators (5.31) through (5.33), the equations of motion for φ^a are

$$\dot{\varphi}^a = i[\widehat{\widetilde{\mathcal{H}}}, \varphi^a].$$

Passing now to the Heisenberg picture in θ and $\bar{\theta}$, we get

$$S\dot{\varphi}^a S^{-1} = iS[\widehat{\widetilde{\mathcal{H}}}, \varphi^a]S^{-1} \implies \dot{\Phi}^a = i[S\widehat{\widetilde{\mathcal{H}}}S^{-1}, \Phi^a] \implies \dot{\Phi}^a = i[\widehat{\widetilde{\mathcal{H}}}, \Phi^a].$$

All the steps above are trivial except the last one, i.e. $S\widehat{\widetilde{\mathcal{H}}}S^{-1} = \widehat{\widetilde{\mathcal{H}}}$, which is due to the fact that both $\widehat{Q}$ and $\widehat{\bar{Q}}$ entering S commute with $\widehat{\widetilde{\mathcal{H}}}$. The beauty of the equation of motion

$$\dot{\Phi}^a = i[\widehat{\widetilde{\mathcal{H}}}, \Phi^a] \qquad (5.68)$$

is that it encapsulates in a single equation all the equations of motion for the $8n$ variables $(\varphi^a, c^a, \bar{c}_a, \lambda_a)$, as can be proved by expanding Eqn.(5.68) in θ, $\bar{\theta}$. We could call Eqn.(5.68) the super-Heisenberg equation of motion.

5.4.1 *Dequantization in the q and p-polarizations and supertime*

In this section we study the role of the superphase space variables Φ^a at the Lagrangian and path integral level.

We have seen that, for what concerns the Hamiltonians, relation (5.67) holds:

$$H[\Phi] = H[\varphi] + \theta N + \bar{N}\bar{\theta} - i\bar{\theta}\theta\widetilde{\mathcal{H}}$$

which implies that

$$i\int \mathrm{d}\theta\mathrm{d}\bar{\theta}H[\Phi] = \widetilde{\mathcal{H}}. \qquad (5.69)$$

An analog of this relation at the Lagrangian level does not hold exactly. The reason, as explained in detail in Appendix F, is the presence in the Lagrangian of the kinetic terms $p\dot{q}$, which are not present in H, i.e.:

$$L(p,q) = p\dot{q} - H(p,q).$$

Replacing q and p in the Lagrangian L with the superphase space variables Φ^q and Φ^p, the analog of Eq. (5.69) becomes the following:

$$i\int d\theta d\bar{\theta} L(\Phi) = \widetilde{\mathcal{L}} - \frac{d}{dt}(\lambda_{p_i}p_i + i\bar{c}_{p_i}c^{p_i}), \tag{5.70}$$

where $\widetilde{\mathcal{L}}$ is the Lagrangian of the CPI given by Eq. (5.15), and the λ_{p_i}, $\bar{c}_{p_i}$ and c^{p_i} are the second half of the variables λ_a, $\bar{c}_a$ and c^a. From now on we will change our notation for the superphase space variables: instead of writing

$$\Phi^a \equiv \varphi^a + \theta c^a + \bar{\theta}\omega^{ab}\bar{c}_b + i\bar{\theta}\theta\omega^{ab}\lambda_b$$

we will explicitly indicate the q and p components in the following manner:

$$\Phi^a = \begin{pmatrix} Q_i \\ P_i \end{pmatrix} \equiv \begin{pmatrix} q_i \\ p_i \end{pmatrix} + \theta\begin{pmatrix} c^{q_i} \\ c^{p_i} \end{pmatrix} + \bar{\theta}\begin{pmatrix} \bar{c}_{p_i} \\ -\bar{c}_{q_i} \end{pmatrix} + i\bar{\theta}\theta\begin{pmatrix} \lambda_{p_i} \\ -\lambda_{q_i} \end{pmatrix}, \tag{5.71}$$

where $i = (1, \cdots, n)$, and $a = (1, \cdots, 2n)$.

Going now back to Eqn.(5.70) we could write it as follows:

$$\widetilde{\mathcal{L}} = i\int d\theta d\bar{\theta} L[\Phi] + \frac{d}{dt}(\lambda_p p + i\bar{c}_p c^p), \tag{5.72}$$

where we have dropped the index "i" appearing on the extended phase space variables. The expression of $\widetilde{\mathcal{L}}$ which appears in Eqn.(5.72) can be used in

$$K(\varphi,c,t|\varphi_0,c_0,t_0) = \int \mathcal{D}''\varphi\,\mathcal{D}\lambda\,\mathcal{D}''c\,\mathcal{D}\bar{c}\,\exp i\int_{t_0}^t d\tau\widetilde{\mathcal{L}}. \tag{5.73}$$

The difference with respect to Eqn.(5.14) is in the measure of integration, which in Eqn.(5.73) has the initial and final c *not* integrated over. Doing so we obtain:

$$\langle\varphi,c|\varphi_0,c_0\rangle = \int \mathcal{D}''\varphi\,\mathcal{D}\lambda\,\mathcal{D}''c\,\mathcal{D}\bar{c}\,\exp i\int_{t_0}^t d\tau\widetilde{\mathcal{L}}$$

$$= \int \mathcal{D}''\varphi\,\mathcal{D}\lambda\,\mathcal{D}''c\,\mathcal{D}\bar{c}\,\exp\left[i\int_{t_0}^t id\tau d\theta d\bar{\theta} L[\Phi] + (\text{s.t.})\right] \tag{5.74}$$

where (s.t.) indicates the surface terms, which come from the total derivative appearing on the r.h.s. of Eqn.(5.72) and has the form

$$(\text{s.t.}) = i\lambda_p p - i\lambda_{p_0}p_0 - \bar{c}_p c^p + \bar{c}_{p_0}c^{p_0}. \tag{5.75}$$

We have indicated with p_0 the n components of the initial momentum. The surface terms present in Eqn.(5.74) somewhat spoil the beauty of formula (5.74) but we can get rid of them by changing the basis of our Hilbert space. At the beginning of Section 5.4 we showed that, besides the basis $|\varphi, c\rangle = |q, p, c^q, c^p\rangle$, we could introduce the "mixed" basis defined in Eqn.(5.57) by the states: $|q, \lambda_p, c^q, \bar{c}_p\rangle$. We can then pass from the transition amplitude $\langle q, p, c^q, c^p, t|q_0, p_0, c^{q_0}, \bar{c}_{p_0}, t_0\rangle$ of Eqn.(5.74) to the mixed one $\langle q, \lambda_p, c^q, \bar{c}_p, t|q_0, \lambda_{p_0}, c^{q_0}, \bar{c}_{p_0}, t_0\rangle$, which are related to each other as follows:

$$\langle q, \lambda_p, c^q, \bar{c}_p, t|q_0, \lambda_{p_0}, c^{q_0}, \bar{c}_{p_0}, t_0\rangle = \qquad (5.76)$$

$$\int dp\, dp_0\, dc^p\, dc^{p_0}\, e^{-i\lambda_p p + \bar{c}_p c^p + i\lambda_{p_0} p_0 - \bar{c}_{p_0} c^{p_0}} \langle q, p, c^q, c^p, t|q_0, p_0, c^{q_0}, c^{p_0}, t_0\rangle.$$

On the r.h.s. of this formula the kernel $\langle q, p, c^q, c^p, t|q_0, p_0, c^{q_0}, c^{p_0}, t_0\rangle$ can be replaced by its path integral expression given in Eqn.(5.74). In this way we get the very *neat* expression:

$$\langle q, \lambda_p, c^q, \bar{c}_p, t|q_0, \lambda_{p_0}, c^{q_0}, \bar{c}_{p_0}, t_0\rangle = \int \mathcal{D}''Q\mathcal{D}P \exp\left[i \int_{t_0}^{t} id\tau d\theta d\bar{\theta} L(\Phi)\right]$$
$$(5.77)$$

where

$$\mathcal{D}''QDP \equiv \mathcal{D}''q\mathcal{D}p\mathcal{D}''\lambda_p \mathcal{D}\lambda_q \mathcal{D}''c^q \mathcal{D}c^p \mathcal{D}''\bar{c}_p \mathcal{D}\bar{c}_q. \qquad (5.78)$$

We would like to point out three "interesting" aspects of Eq. (5.77).

1. Note that the surface terms of Eqn.(5.74) have disappeared in Eqn.(5.77);

2. The measure in the path integral (5.77) is the same as the measure of the quantum path integral, which is

$$\langle q, t|q_0, t_0\rangle = \int \mathcal{D}''q\mathcal{D}p \, \exp\frac{i}{\hbar} \int d\tau\, L[\varphi] \qquad (5.79)$$

but with p replaced by P and q by Q. Both in Eqn.(5.77) and in Eqn.(5.79) the integration in p and P is done even over the initial and final variables while the integration in q and Q is done only over the intermediate points between the initial and final ones. The reason for the notation (5.78) should be clear from the fact that the superphase space variables Q, P are defined as

$$Q \equiv q + \theta c^q + \bar{\theta}\bar{c}_p + i\bar{\theta}\theta\lambda_p$$
$$P \equiv p + \theta c^p - \bar{\theta}\bar{c}_q - i\bar{\theta}\theta\lambda_q,$$
$$(5.80)$$

so the integration over Q, P means the integration over the elements $(q, c^q, \bar{c}_p, \lambda_p)$ and $(p, c^p, \bar{c}_q, \lambda_q)$ which make up Q and P respectively;

3. The function L that enters both the QM path integral (5.79) and the CM one (5.77) is the same. The only difference is that in QM (5.79) the variables entering L are the normal phase space variables while in CM (5.77) they are the superphase space variables $\Phi^a = (Q, P)$.

Let us now proceed by noticing that formally we can rewrite Eqn.(5.77) as

$$\langle Q, t | Q_0, t_0 \rangle \equiv \int \mathcal{D}'' Q \mathcal{D}P \exp\left[i \int_{t_0}^{t} id\tau d\theta d\bar{\theta} L[\Phi]\right] \qquad (5.81)$$

where we have defined the ket $|Q\rangle$ as the common eigenstate of the operators $\hat{q}$, $\hat{\lambda}_p$, $\hat{c}^q$, $\hat{\bar{c}}_p$:

$$\hat{q}|Q\rangle = q|Q\rangle, \qquad\qquad \hat{\lambda}_p|Q\rangle = \lambda_p|Q\rangle,$$
$$\hat{c}^q|Q\rangle = c^q|Q\rangle, \qquad\qquad \hat{\bar{c}}_p|Q\rangle = \bar{c}_p|Q\rangle. \qquad (5.82)$$

So $|Q\rangle$ can be "identified" with the state $|q, \lambda_p, c^q, \bar{c}_p\rangle$ which appears in Eqn.(5.77). The reader may not like this notation because in Eqn.(5.80) Q contains the Grassmann variables θ and $\bar{\theta}$ which do not appear at all in Eqn.(5.82). Actually, from Eqn.(5.82) we can derive that the state $|Q\rangle$ is also an eigenstate of the supervariable $\hat{Q}$ obtained by turning the expression (5.80) into an operator, i.e.:

$$\hat{Q}(\theta, \bar{\theta})|Q\rangle = Q(\theta, \bar{\theta})|Q\rangle. \qquad (5.83)$$

This relation is just a simple consequence of Eqn.(5.82) as can be proved by expanding in θ and $\bar{\theta}$ both $\hat{Q}$ and Q in Eqn.(5.83). One immediately sees that the variables θ and $\bar{\theta}$ make their appearance not in the state $|Q\rangle$ but in its eigenvalue Q and in the operator $\hat{Q}$. We are now ready to compare Eqns.(5.79) and (5.81). One is basically the central element of QM:

$$\langle q, t | q_0, t_0 \rangle = \int \mathcal{D}'' q \mathcal{D}p \exp\left[\frac{i}{\hbar} \int_{t_0}^{t} d\tau L[\varphi]\right], \qquad (5.84)$$

while the other is the central element of CM (formulated à la KvN or à la CPI):

$$\langle Q, t | Q_0, t_0 \rangle = \int \mathcal{D}'' Q \mathcal{D}P \exp\left[i \int_{t_0}^{t} id\tau d\theta d\bar{\theta} \, L[\Phi]\right]. \qquad (5.85)$$

By just looking at Eqns.(5.84) and (5.85), it is now easy to give some simple *rules*, which turn the *quantum* transition amplitude (5.84) into the *classical* one (5.85). The rules are:

1) *Replace in the QM case the phase space variables* (q, p) *everywhere with the superphase space ones* (Q, P);

2) *Extend the time integration to the supertime integration multiplied by* $\hbar$

$$\int d\tau \longrightarrow i\hbar \int d\tau d\theta d\bar{\theta}. \qquad (5.86)$$

The reason for the appearance of the "i" on the r.h.s. of Eqn.(5.86) is related to the complex nature of the Grassmann variables θ and $\bar{\theta}$, as explained in Appendix E. The reason for the appearance of $\hbar$ instead is related to the fact that in Eqn.(5.85), which is CM, there is no $\hbar$ and so in Eqn.(5.86) we need to introduce an $\hbar$ in order to cancel the one of Eqn.(5.84). From the dimensional point of view formula (5.86) is correct because, as shown in Appendix E, the dimensions of $d\theta\, d\bar{\theta}$ are just the inverse of an action canceling in this manner the dimension of $\hbar$ appearing in front of the r.h.s. in Eqn.(5.86). This implies that both the l.h.s. and the r.h.s. of (5.86) have the dimension of a time. We will indicate the rules 1) and 2) above as dequantization rules.

Note that these *dequantization* rules are *not* the semiclassical or WKB limit of QM. In fact we are not sending $\hbar \to 0$ in Eqn.(5.84) and what we get is not QM in the leading order in $\hbar$, like in the WKB method, but exactly CM in the KvN or CPI formulation. We named this procedure "dequantization" because it is the inverse of "quantization" in the sense that while quantization is a set of rules which turn CM into QM, our rules 1) and 2) turn QM into CM. We called our procedure "*geometrical*" because it basically consists of a geometrical *extension* of both the base space given by time t, into the supertime $(t, \theta, \bar{\theta})$, and of the target space, which is phase space (q, p) in QM, into a superphase space (Q, P) in CM. We used the expressions "*base space*" and "*target space*", as it is done nowadays in strings and higher dimensional theories, where procedures of dimensional extension or dimensional contraction are very often encountered. In those theories the procedures of dimensional extension is introduced to give a geometrical basis to the many extra fields present in grand-unified theories, while the procedure of dimensional contraction is needed to come back to our four-dimensional world. We find it amazing and thought-provoking that even the procedure of quantization (or dequantization) can be achieved via a dimensional contraction (or extension).

In [Abrikosov Jr. and Gozzi (2000); Abrikosov Jr. *et al.* (2003)] we have given brief presentations of these ideas but there we explored the inverse route, that is the one of *quantization*, which is basically how to pass from Eqn.(5.85) to Eqn.(5.84). This goal is achieved by a sort of dimensional reduction from the supertime $(t, \theta, \bar{\theta})$ to the time t, and from the super-phase space (Q, P) to the phase space (q, p). In those papers [Abrikosov Jr. and Gozzi (2000); Abrikosov Jr. *et al.* (2003)] we thought of implementing the supertime reduction by inserting a $\delta(\bar{\theta})\delta(\theta)/\hbar$ into the weight appearing in Eqn.(5.85) but we found this method a little bit awkward and that is why here we have preferred to explore the opposite route that is the one of *dequantization* which is brought about by a dimensional extension. Even if awkward to implement, the quantization route from Eqn.(5.85) to Eqn.(5.84) can be compared with a method of quantization which we mentioned before and known in the literature as *geometric quantization* (GQ) [Woodhouse (1997)]. We will not review it here but suffice to say that it starts from the so-called "prequantization space" (which is our space of KvN states $\psi(q, p)$), and from the *Lie derivative of the Hamiltonian flow* (which is our $\widetilde{\mathcal{H}}$ of Eqn.(5.25)), and, through a long set of steps, it builds up the Schrödinger operator and the Hilbert space of QM. Basically, in going from Eqn.(5.85) to Eqn.(5.84), we do the same because we go from the weight $\exp i \int i \, dt d\theta d\bar{\theta} L[\Phi]$, which is the evolution via the Lie derivative operator $\exp(-i\widetilde{\mathcal{H}}t)$, to the weight $\exp \frac{i}{\hbar} \int L[\varphi]$, which reproduces the evolution via the Schrödinger operator $\exp(-i\frac{\hat{H}}{\hbar}t)$. Regarding the states we go from the KvN states $|Q\rangle$ to the Schrödinger ones $|q\rangle$ by just sending $\theta, \bar{\theta} \to 0$ in Eqn.(5.83). The difference with respect to GQ is that our KvN states contain also the Grassmann variables c and $\bar{c}$, which are not contained in the prequantization states of GQ. In GQ the reduction of the KvN states to the quantum ones is achieved via a procedure called "polarization" while the transformation of the Lie derivative into the quantum Schrödinger operator is achieved via a totally different procedure, see [Woodhouse (1997)] for details. In our opinion it is somewhat unpleasant that in GQ states and operators are "quantized" via two totally different procedures. This is not so anymore in our functional approach, which brings Eqn.(5.85) in Eqn.(5.84). It is in fact the dimensional reduction both in t

$$i\hbar \int dt d\theta d\bar{\theta} \longrightarrow \int dt \qquad (5.87)$$

and in phase space

$$(Q, P) \longrightarrow (q, p), \qquad (5.88)$$

which at the same time produces the right operators (from the classical Lie derivative to the Schrödinger operator) and the right states (from the KvN states $|Q\rangle$ to the quantum ones $|q\rangle$). So we do not need two different procedures for operators and states but just a single one. Actually the dimensional reduction contained in Eqns.(5.87) and (5.88) can be combined into a single operation, that is the one of *shrinking to zero* the variables $\theta, \bar{\theta}$, i.e.: $(\theta, \bar{\theta}) \to 0$. This not only brings the integration $\int dt d\theta d\bar{\theta}$ to $\int dt$ but, remembering the form of (Q, P) i.e. Eqn.(5.80), it also brings

$$Q \longrightarrow q.$$

Because of this, it reduces the KvN states $|Q\rangle$ to the quantum ones $|q\rangle$, which are a basis for the quantum Hilbert space in the Schrödinger representation. Note the difference with the GQ procedure: there one starts with the states $|q, p\rangle$ and the "p" is removed through a long set of steps, known as polarization [Woodhouse (1997)]. In our approach instead we first replace the "p" in the $|q, p\rangle$ states with the λ_p via the *Fourier transform* presented in Eqn.(5.76) and then remove the λ_p by *sending* $(\theta, \bar{\theta}) \to 0$. We want to stress again that the same two steps, 1) *Fourier transform* and 2) *sending* $\theta, \bar{\theta} \to 0$, which polarize the states, are the same ones which turn the *classical* evolution into the *quantum* one. In fact step 1) takes away the surface terms in Eqn.(5.74) bringing the weight to be of the same form as the quantum one and step 2), sending $(\theta, \bar{\theta}) \to 0$, brings the classical weight to the quantum one. We feel that this coincidence of the two procedures, for the states and the operators, was not noticed in GQ because there they did not use the partners of time $(\theta, \bar{\theta})$ and the functional approach. This coincidence is quite interesting because (besides a trivial Fourier transform) it boils down to be a *geometrical* operation: the dimensional reduction from supertime $(t, \theta, \bar{\theta})$ to time t. We feel that this is really the *geometry* at the heart of geometric quantization and of quantum mechanics in general.

The reader anyhow may suspect that the simple operation of sending $\theta, \bar{\theta} \to 0$ may well produce both the right quantum evolution and at the same time polarize the states but that this coincidence takes place only in the Schrödinger polarization. We will prove that this is not so. Below we shall show that all this procedure works also in the momentum polarization and in [Abrikosov Jr. *et al.* (2005)] the reader can find also the coherent states construction.

Analogously to what we did in Eqn.(5.76) let us perform a partial Fourier transform in order to go from the standard $\langle q, p, c^q, c^p|$ basis of the CPI to the $\langle \lambda_q, p, \bar{c}_q, c^p|$ one. In this new basis the transition amplitude is related to the old one in the following manner:

$$\langle \lambda_q, p, \bar{c}_q, c^p, t | \lambda_{q_0}, p_0, \bar{c}_{q_0}, c^{p_0}, t_0 \rangle \tag{5.89}$$

$$= \int dq dq_0 dc^q dc^{q_0} e^{-i\lambda_q q + \bar{c}_q c^q} \langle q, p, c^q, c^p, t | q_0, p_0, c^{q_0}, c^{p_0}, t_0 \rangle e^{i\lambda_{q_0} q_0 - \bar{c}_{q_0} c^{q_0}}.$$

Replacing the kernel $\langle q, p, c^q, c^p, t | q_0, p_0, c^{q_0}, c^{p_0}, t_0 \rangle$ on the r.h.s. of the formula above with its path integral expression given in Eqn.(5.74) and using a properly defined discretized form of this path integral, we get:

$$\langle \lambda_q, p, \bar{c}_q, c^p, t | \lambda_{q_0}, p_0, \bar{c}_{q_0}, c^{p_0}, t_0 \rangle \tag{5.90}$$

$$= \int \mathcal{D}Q \mathcal{D}'' P \exp i \int_{t_0}^{t} i d\tau d\theta d\bar{\theta} \left\{ L[\Phi] - \frac{d(QP)}{d\tau} \right\}.$$

Different from the case of the $\langle q, \lambda_p, c^q, \bar{c}_q |$ states, the surface terms coming from the partial Fourier transform in Eqn.(5.89) and those coming from Eqn.(5.75) do not cancel against each other. Nevertheless, as indicated explicitly, the surface terms that remain in Eqn.(5.90), i.e. $\dfrac{d(QP)}{dt}$, can be written in terms of the superphase space variables Q and P, as we will prove in detail in Appendix G. Also the states appearing in Eqn.(5.90), i.e. $|\lambda_q, p, \bar{c}_q, c^p \rangle$, can be formally written in terms of the supervariables. In fact we should note that the variables $(\lambda_q, p, \bar{c}_q, c^p)$ entering these states are exactly the components of the P of Eqn.(5.80). Analogously to what we did in Eqn.(5.82) for the state $|Q\rangle$, we could define the state $|P\rangle$ as the common eigenstate of the operators $\hat{p}, \widehat{\lambda}_q, \widehat{\bar{c}}_q$ and $\widehat{c}^p$:

$$\hat{p}|P\rangle = p|P\rangle, \qquad \widehat{\lambda}_q |P\rangle = \lambda_q |P\rangle,$$

$$\widehat{\bar{c}}_q |P\rangle = \bar{c}_q |P\rangle, \qquad \widehat{c}^p |P\rangle = c^p |P\rangle.$$

The equations above can also be written in a compact form as $\widehat{P}(\theta, \bar{\theta}) |P\rangle = P(\theta, \bar{\theta}) |P\rangle$. It is then natural to identify the state $|P\rangle$ with the state $|\lambda_q, p, \bar{c}_q, c^p \rangle$ appearing in Eqn.(5.90)[2], which can be rewritten in the very compact form:

$$\langle P, t | P_0, t_0 \rangle = \int \mathcal{D}Q \mathcal{D}'' P \exp i \int_{t_0}^{t} i d\tau d\theta d\bar{\theta} \left\{ L[\Phi] - \frac{d(QP)}{d\tau} \right\}. \tag{5.91}$$

---

[2]Using the completeness relations which arise from these $|P\rangle$ states and the $|Q\rangle$ ones of Eqn.(5.83) it is possible to build, via the standard Trotter formula, the path integral of the CPI directly in the superfield (or superphase variables) form. It is also possible to derive the graded commutators indicated in Eqns.(5.31) though (5.33) via generalized commutation relations among the superfields. Further details can be found in Appendix H.

Let us now recall the expression of the *quantum* transition amplitude in the momentum representation

$$\langle p,t|p_0,t_0\rangle = \int dqdq_0\, e^{-ipq/\hbar}\langle q,t|q_0,t_0\rangle e^{+ip_0q_0/\hbar}. \tag{5.92}$$

By using a properly defined discretized form we get for Eqn.(5.92) the following expression[3]

$$\langle p,t|p_0,t_0\rangle = \int \mathcal{D}q\mathcal{D}''p \,\exp\frac{i}{\hbar}\int_{t_0}^{t} d\tau\left(L(q,p)-\frac{d(qp)}{d\tau}\right). \tag{5.93}$$

At this point, looking at the *quantum* expression (5.93) and at the *classical* one (5.91), it is clear that the dequantization procedure that leads us from Eqn.(5.93) to Eqn.(5.91) consists of the same two rules which worked in the coordinate representation and which we wrote in italics in the lines below Eqn.(5.85). This confirms that our procedure works also in the momentum representation and that also in this case the same set of rules produces both the right evolution operator and the right representation.

Before concluding this section, we would like to draw again the reader's attention to three crucial things, which made the whole procedure work as nicely as it did. The *first* one is that the classical weight $\int dt\,\widetilde{\mathcal{L}}$ and the quantum one $\int dt\,L$ belong, modulo surface terms, to the same multiplet. In fact if we expand $S[\Phi] = \int dtL[\Phi]$ in θ and $\bar\theta$ we get:

$$S[\Phi] = \int dtL(\varphi) + \theta\mathcal{T}(\varphi,\lambda,c,\bar c) + \bar\theta\mathcal{V}(\varphi,\lambda,c,\bar c)$$

$$+i\theta\bar\theta\left(\int dt\,\widetilde{\mathcal{L}}(\varphi,\lambda,c,\bar c) + \text{s.t.}\right), \tag{5.94}$$

where the functions $\mathcal{T}$ and $\mathcal{V}$ are the analog of the N and $\bar N$ which appeared in Eqn.(5.66). It is not important to write down the explicit form of $\mathcal{T}$ and $\mathcal{V}$ but to note that in Eqn.(5.94) the weights entering respectively the QPI, i.e. $\int dtL(\varphi)$, and the CPI, i.e. $\int dt\widetilde{\mathcal{L}}(\varphi,\lambda,c,\bar c)$, belong to the same multiplet (modulo surface terms). Somehow we can say that *"God has put CM and QM in the same multiplet (modulo surface terms)"*.

The *second thing* to note is that in the statement in italics and between quotation marks written above we can even drop the "modulo surface terms" part of the sentence. In fact those surface terms are crucial because, combined with the extra surface terms coming from the partial

---

[3]To make the reader understand how one can pass from the $\mathcal{D}''q\mathcal{D}p$ measure to the $\mathcal{D}q\mathcal{D}''p$ one in Appendix G we report the discretized derivation of this formula which can anyhow be found in the literature, see for example [Kleinert (1990)].

Fourier transforms of Eqn.(5.76) or (5.89), they give exactly the classical weight that goes into the quantum one by the single procedure of sending $\theta, \bar{\theta} \to 0$.

The *third crucial thing* to note is that the partial Fourier transform of Eqn.(5.76) or (5.89), which produces the extra surface terms exactly needed to implement the procedure above, produces also the right classical states that go into the quantum ones by the same process of sending $\theta, \bar{\theta} \to 0$. This set of "incredible coincidences" works both for the momentum and the coordinate representations and, as shown in [Abrikosov Jr. *et al.* (2005)], also for the coherent states ones.

The last issue that we want to touch on is the one of ordering ambiguities. We know that there are no ordering problems in CM, at least in its standard formulation, while they are present in QM. This issue is at the basis of the fact that the quantization is not unique, i.e. starting from the same classical Hamiltonian we can end up in many different ones at the quantum level according to the ordering we choose. Is this ambiguity resolved by our method of quantization? The answer is no. In fact CM, even if formulated à la KvN (or CPI), presents no ordering ambiguities as proved in detail in Appendix I. A detailed proof of this fact was needed because in principle the KvN (and CPI) formulation contains operators like $\widehat{\varphi}^a$, $\widehat{\lambda}_a$, which do not commute and so it could have presented ordering ambiguities. Having no ordering problem at the CPI level we conclude that these problems are injected into QM, not by the non-commuting operators of the CPI, but by the limiting procedure of sending $\theta, \bar{\theta} \to 0$. This limit in fact changes the kinetic term in the CPI from $\Phi^p \dot{\Phi}^q$ to $p\dot{q}$ which in turn implies the non-commutativity of p and q at the quantum level. We can conclude by saying that our quantization procedure does not solve the lack of uniqueness present in the standard quantization process.

5.4.2   *Generating functionals and Dyson-Schwinger equations*

The correspondence that we established between the transition amplitudes (5.84) and (5.85) can be extended also to more general objects, like the "analog" of the generating functionals $Z[J]$ of quantum field theory. The object we have in mind can be defined as follows

$$Z_{\mathrm{QM}}\left([J_a(t)]; q, q_0, t - t_0\right) = \int_{q_0}^{q} \mathcal{D}'' q \mathcal{D} p \, \exp\left[\frac{i}{\hbar} \int_{t_0}^{t} d\tau \, \left(L[\varphi] + J_a \varphi^a\right)\right].$$

$$(5.95)$$

It is not the vacuum-vacuum functional of quantum field theory but a *functional* of some external currents J_a and a *function* of the initial (q_0, t_0) and final (q, t) points which generalizes the transition amplitude (5.84).

The classical analog of Z_{QM} is the generalization at the level of generating functional [Gozzi *et al.* (1989)] of the expression (5.85):

$$Z_{\mathrm{CM}}\left([\mathbb{J}_a(t, \theta, \bar{\theta})]; Q, Q_0, t - t_0\right) \qquad (5.96)$$

$$= \int_{Q_0}^{Q} \mathcal{D}''Q\mathcal{D}P \exp\left[i \int_{t_0}^{t} i \mathrm{d}\tau\mathrm{d}\theta\mathrm{d}\bar{\theta} \ L[\Phi]\right.$$

$$\left. +i \int_{t_0}^{t} \mathrm{d}\tau \left(J_{\varphi^a}\varphi^a + J_{\lambda_a}\lambda_a + J_{c^a}c^a + \bar{c}_a J_{\bar{c}_a}\right)\right].$$

The $8n$ currents $(J_{\varphi^a}, J_{\lambda_a}, J_{c^a}, J_{\bar{c}_a})$ can be grouped together in a supercurrent $\mathbb{J}_a$ defined as

$$\mathbb{J}_a \equiv \omega_{ab}J_{\lambda_b} - i\theta\omega_{ab}J_{\bar{c}_b} + i\bar{\theta}J_{c^a} - i\bar{\theta}\theta J_{\varphi^a}.$$

Using this expression, the terms of Eqn.(5.96) depending on the currents can be written in a more compact form as

$$\int_{t_0}^{t} \mathrm{d}\tau \left(J_{\varphi^a}\varphi^a + J_{\lambda_a}\lambda_a + J_{c^a}c^a + \bar{c}_a J_{\bar{c}_a}\right) = i \int_{t_0}^{t} \mathrm{d}\tau\mathrm{d}\theta\mathrm{d}\bar{\theta}(\mathbb{J}_a\Phi^a).$$

In this manner the Z_{CM} of Eqn.(5.96) assumes the following compact expression

$$Z_{\mathrm{CM}}\left([\mathbb{J}_a(t, \theta, \bar{\theta})]; Q, Q_0, t - t_0\right)$$

$$= \int_{Q_0}^{Q} \mathcal{D}''Q\mathcal{D}P \exp\left[i \int_{t_0}^{t} i \mathrm{d}\tau\mathrm{d}\theta\mathrm{d}\bar{\theta} \ \{L[\Phi] + \mathbb{J}_a\Phi^a\}\right], \qquad (5.97)$$

which can be compared to the quantum one of Eqn.(5.95)

$$Z_{\mathrm{QM}}([J_a(t)]; q, q_0, t - t_0) = \int_{q_0}^{q} \mathcal{D}''q\mathcal{D}p \exp\left[\frac{i}{\hbar} \int_{t_0}^{t} \mathrm{d}\tau(L[\varphi] + J_a\varphi^a)\right]. \qquad (5.98)$$

One immediately notices that the dequantization, i.e. passing from Z_{QM} to Z_{CM}, is achieved by applying to the r.h.s. of Eqn.(5.98) the following three rules

1) $\int \mathrm{d}\tau \longrightarrow i\hbar \int \mathrm{d}\tau\mathrm{d}\theta\mathrm{d}\bar{\theta}$,

2) $\varphi^a \longrightarrow \Phi^a$,

3) $J_a(t) \longrightarrow \mathbb{J}_a(t, \theta, \bar{\theta})$.

There is the rule no. **3)**, besides the two old ones, because Z is a functional of the current. It is clear that these rules work because Z_{CM} and Z_{QM} have the same functional form. This identity in the functional form has a further consequence. We know in fact that the generating functional $Z[J]$ satisfies a functional equation, known as Dyson-Schwinger equation (see for example [Zinn-Justin (1996)]). Since both Z_{CM} and Z_{QM} have the same functional form we expect that they satisfy two Dyson-Schwinger equations with the same functional form. This is what we are going to prove.

We know [Zinn-Justin (1996)] that the integral of a total derivative is zero and this works also in the functional case, so once these two operations are applied in sequence to a functional $F[\varphi]$ we get zero, i.e.

$$\int \mathcal{D}\varphi \frac{\delta}{\delta \varphi} F[\varphi] = 0.$$

If we choose $F[\varphi]$ to be

$$\exp\left[\frac{i}{\hbar} \left(S[\varphi] + \int dt J_a \varphi^a \right) \right]$$

where $S[\varphi] = \int dt L(\varphi)$, we get:

$$\int \mathcal{D}\varphi \frac{\delta}{\delta \varphi} \left[\exp \frac{i}{\hbar} \left(S[\varphi] + \int dt J_a \varphi^a \right) \right] = 0,$$

which is equivalent to:

$$\int \mathcal{D}\varphi \left[\frac{\delta S}{\delta \varphi} + J \right] \exp\left[\frac{i}{\hbar} \left(S[\varphi] + \int dt J_a \varphi^a \right) \right] = 0. \qquad (5.99)$$

Remembering that

$$-i\hbar \frac{\delta}{\delta J_a} \exp\left[\frac{i}{\hbar} \int dt J_a \varphi^a \right] = \varphi^a \exp\left[\frac{i}{\hbar} \int dt J_a \varphi^a \right] \qquad (5.100)$$

we can replace the fields φ^a with $-i\hbar \dfrac{\delta}{\delta J_a}$ and rewrite Eqn.(5.99) as:

$$\left[\frac{\delta S}{\delta \varphi} \left(-i\hbar \frac{\delta}{\delta J} \right) + J(t) \right] Z_{\text{QM}}[J] = 0, \qquad (5.101)$$

where we have used the explicit expression of the generating functional (5.95). Eqn.(5.101) is the Dyson-Schwinger equation satisfied by the generating functional $Z_{\text{QM}}[J]$ and it can also be rewritten as:

$$\left[\partial_t \left(-i\hbar \frac{\delta}{\delta J_a(t)} \right) - \omega^{ab} \partial_b H \left(-i\hbar \frac{\delta}{\delta J(t)} \right) + \omega^{ab} J_b \right] Z_{\text{QM}}[J_a] = 0, \qquad (5.102)$$

where the combination of functional derivatives present on the l.h.s. of Eqn.(5.102) comes from the term $\dfrac{\delta S}{\delta \varphi}$ in Eqn.(5.101), which in our case is $\partial_t \varphi^a - \omega^{ab} \partial_b H(\varphi)$, and by replacing φ with $-i\hbar \dfrac{\delta}{\delta J}$ as dictated by the rule (5.100) of the Dyson-Schwinger equation.

Performing the same steps for the *classical* generating functional $Z_{\mathrm{CM}}[\mathbb{J}]$ written in Eqn.(5.97) we get

$$\left[\partial_t \left(-\frac{\delta}{\delta \mathbb{J}_a(t,\theta,\bar\theta)} \right) - \omega^{ab} \partial_b H \left(-\frac{\delta}{\delta \mathbb{J}(t,\theta,\bar\theta)} \right) + \omega^{ab} \mathbb{J}_b(t,\theta,\bar\theta) \right] Z_{\mathrm{CM}}[\mathbb{J}] = 0.$$
(5.103)

The combination of superfunctional derivatives present on the l.h.s. of Eqn.(5.103) is obtained from the l.h.s. of the super-equations of motion: $\partial_t \Phi^a - \omega^{ab} \dfrac{\partial H}{\partial \Phi^b}(\Phi) = 0$ which are formally equivalent to the set of $8n$ equations of motion for $(\varphi^a, \lambda_a, c^a, \bar c_a)$. To get Eqn.(5.103) we must also remember that the definition of the functional derivative with respect to the supercurrent, $\dfrac{\delta}{\delta \mathbb{J}_a(t,\theta,\bar\theta)} \mathbb{J}_d(t,\theta',\bar\theta') = \delta^a_d \delta(\bar\theta'-\bar\theta)\delta(\theta'-\theta)$, implies that the analog of Eqn.(5.100) is given by:

$$\frac{\delta}{\delta \mathbb{J}_a(t,\theta,\bar\theta)} \left[\exp\left(i \int i dt' d\theta' d\bar\theta' \, \mathbb{J}_d(t',\theta',\bar\theta') \Phi^d(t',\theta',\bar\theta') \right) \right]$$
$$= -\Phi^a(t,\theta,\bar\theta) \left[\exp\left(i \int i dt' d\theta' d\bar\theta' \, \mathbb{J}_d(t',\theta',\bar\theta') \Phi^d(t',\theta',\bar\theta') \right) \right].$$

The previous equation tells us that we can replace the superfield variable $\Phi(t,\theta,\bar\theta)$ with the functional derivative $-\dfrac{\delta}{\delta \mathbb{J}(t,\theta,\bar\theta)}$, as we did in order to get Eqn.(5.103).

Eqns.(5.103) and (5.102) prove that the *classical* and the *quantum* Dyson-Schwinger equations are formally very similar. One can pass from one to the other via the replacements

1) $J_a(t) \longrightarrow \mathbb{J}_a(t,\theta,\bar\theta)$,

2) $-i\hbar \dfrac{\delta}{\delta J_a(t)} \longrightarrow -\dfrac{\delta}{\delta \mathbb{J}_a(t,\theta,\bar\theta)}$.

The reader may think that rules 1) and 2) are in contradiction with each other because in 1) there is no $\hbar$ while in 2) an $\hbar$ makes its appearance. Actually there is no contradiction because we should remember that the functional derivative acts on integrated things, so for example $\dfrac{\delta}{\delta \mathbb{J}_a(t,\theta,\bar\theta)}$

acts on objects like $\int dt d\theta d\bar{\theta} \, F[\mathbb{J}_a(t, \theta, \bar{\theta})]$. We know that one of the dequantization rules is

$$\int dt \longrightarrow i\hbar \int dt d\theta d\bar{\theta}.$$

So if we pass from the functional derivation in $J(t)$ to the extended one:

$$\frac{\delta \int dt' F[J(t')]}{\delta J(t)} \longrightarrow \frac{\delta \int dt' d\theta' d\bar{\theta}' F[\mathbb{J}(t', \theta', \bar{\theta}')]}{\delta \mathbb{J}(t, \theta, \bar{\theta})}$$

$$= \frac{\delta \left[i\hbar \int dt' d\theta' d\bar{\theta}' F[\mathbb{J}(t', \theta', \bar{\theta}')] \right]}{i\hbar \, \delta \mathbb{J}(t, \theta, \bar{\theta})}$$

we see that the correspondence between the functional derivatives is $\dfrac{\delta}{\delta J(t)} \longrightarrow \dfrac{1}{i\hbar} \dfrac{\delta}{\delta \mathbb{J}(t, \theta, \bar{\theta})}$ which is equivalent to rule **2)**:

$$-i\hbar \frac{\delta}{\delta J(t)} \longrightarrow -\frac{\delta}{\delta \mathbb{J}(t, \theta, \bar{\theta})}.$$

5.4.3 *Warnings on the dequantization rules*

The three dequantization rules that we have proposed up to now, and listed under Eqn.(5.98), need some further specifications that we are going to discuss in what follows.

A) If we take the path integral expression for the quantum transition amplitude (5.84) and perform explicitly the functional integration we will get a function $K_{\mathrm{QM}}(q, q_0; t - t_0)$ of the initial and final configurations q_0 and q, and of the interval of time $(t - t_0)$:

$$\langle q, t | q_0, t_0 \rangle = K_{\mathrm{QM}}(q, q_0; t - t_0).$$

Analogously if we do the same for the classical transition amplitude (5.85) we will obtain something like:

$$\langle Q, t | Q_0, t_0 \rangle = \widetilde{K}_{\mathrm{CM}}(q, \lambda_p, c^q, \bar{c}_p, q_0, \lambda_{p_0}, c^{q_0}, \bar{c}_{p_0}; t - t_0),$$

where $\widetilde{K}_{\mathrm{CM}}$ is a function of the components of the initial and final superfields Q_0 and Q. If we now naively apply one of the dequantization rules which says: "Replace q with Q in order to pass from QM to CM", we expect that

$$K_{\mathrm{QM}}(Q, Q_0; t - t_0) = \widetilde{K}_{\mathrm{CM}}. \tag{5.104}$$

This is not so for a simple reason: the classical transition amplitude $\tilde{K}_{\mathrm{CM}}$ on the r.h.s. of Eqn.(5.104) is equivalent to the kernel
of propagation (5.76), which does not depend on the Grassmann
partners of time θ and $\bar{\theta}$, while the $K_{\mathrm{QM}}(Q, Q_0)$ on the l.h.s. of
Eqn.(5.104) would depend on θ, $\bar{\theta}$ via the supervariables Q and
Q_0.

This is a first indication that we should not apply the dequantization rule mentioned above in an indiscriminate way. The replacement $q \to Q$ or $\varphi \to \Phi$ works only if we apply it to the functional
integration measure and to the weight entering the path integral,
or in general to any *functional* expression, like it happens for the
Dyson-Schwinger equation. The replacement does not work if we
do it in a *function*.

B) A further example of this fact comes from the comparison between
the quantum commutators

$$\left[\widehat{\varphi}^a(t), \widehat{\varphi}^b(t) \right] = i\hbar\omega^{ab} \qquad (5.105)$$

and the classical ones given by Eqns.(5.31)) through (5.33) which,
as shown in Appendix H, can be easily written in terms of superphase space variables as

$$\left[\widehat{\Phi}^a(t, \theta, \bar{\theta}), \widehat{\Phi}^b(t, \theta', \bar{\theta}') \right] = \omega^{ab}\delta(\bar{\theta} - \bar{\theta}')\delta(\theta - \theta'). \qquad (5.106)$$

We immediately see that, replacing in Eqn.(5.105) $\widehat{\varphi}^a$ with $\widehat{\Phi}^a$, we
do not end up in Eqn.(5.106). So the simple dequantization rule,
which says "replace φ with Φ", does not work also in this case.
Again we can apply it, but at a *functional* level. So we should start
from the path integral (or functional) expression that is responsible
for producing, at the operatorial level, the relation (5.105). That
expression is the kinetic part of the quantum path integral (5.84)

$$\int \mathcal{D}''q \mathcal{D}p \exp \frac{i}{\hbar} \int_{t_0}^{t} \mathrm{d}\tau \left[p\dot{q} - (\cdots) \right].$$

Applying the dequantization rules: $\varphi^a \to \Phi^a$ and $\int \mathrm{d}\tau \to$
$i\hbar \int \mathrm{d}\tau \mathrm{d}\theta \mathrm{d}\bar{\theta}$ at this *functional* level, we get

$$\int \mathcal{D}''QDP \exp i \int_{t_0}^{t} i\mathrm{d}\tau \mathrm{d}\theta \mathrm{d}\bar{\theta} \left[P\dot{Q} - (\cdots) \right]$$

and from this expression we can derive, as proved in Appendix
H, the commutators (5.106). So we can pass from Eqn.(5.105) to

Eqn.(5.106), but not by naively doing the replacement $\varphi^a \to \Phi^a$ at the level of the *functions* in Eqn.(5.105). We have to do that replacement, as in the case **A)** analyzed before, at the level of the path integral (or *functional*) level which produces the expression (5.105).

C) A third example, which tells us that we have to apply the dequantization rules only at the *functional* level, comes from looking at the observables of the quantum theory. We know that in QM the observables are given by Hermitian operators which are functions of $\hat{p}$ and $\hat{q}$. Let us indicate them as $O(\widehat{\varphi}^a)$:

$$[O(\widehat{\varphi}^a)]^\dagger = O(\widehat{\varphi}^a).$$

If we naively apply the dequantization rule $\widehat{\varphi}^a \to \widehat{\Phi}^a$ we would get that the classical observables are given by $O(\widehat{\Phi}^a)$. These are objects that would depend on θ and $\bar{\theta}$, while we know that the classical observables are just functions of p and q and the generators of the canonical transformations at the CPI level are the Lie derivatives along the Hamiltonian vector fields associated with $O(\varphi^a)$. How can we recover these objects by going through our dequantization rules? Again the trick is to apply the rules at the *functional* level. How can we go from a quantum observable to some *"functional"* expression? Well, we can associate to every Hermitian operator $O(\widehat{\varphi})$ a unitary one, defined as

$$\widehat{U}_O(\alpha) \equiv \exp\left[-\frac{i}{\hbar} O(\widehat{\varphi})\alpha\right],$$

where α is a real parameter labeling the transformation generated by $O(\widehat{\varphi})$. We can then sandwich this operator among two states $\langle q|$ and $|q_0\rangle$:

$$\langle q| \exp\left[-\frac{i}{\hbar} O(\widehat{\varphi})\alpha\right] |q_0\rangle. \tag{5.107}$$

This expression is analogous to the kernel of evolution where $O(\widehat{\varphi})$ is replaced by $H(\widehat{\varphi})$ and α is the interval of time t

$$\langle q| \exp\left[-\frac{i}{\hbar} H(\widehat{\varphi})t\right] |q_0\rangle. \tag{5.108}$$

We know that Eqn.(5.108) has a path integral expression obtained by slicing the interval t in N parts and inserting the completeness relations. The same can be done for Eqn.(5.107) as follows:

$$\langle q| \exp\left[-\frac{i}{\hbar}O(\widehat{\varphi})\alpha\right]|q_0\rangle$$

$$= \lim_{N\to\infty}\langle q|e^{-\frac{i}{\hbar}\widehat{O}\frac{\alpha}{N}}\int\frac{\mathrm{d}p_N}{2\pi\hbar}|p_N\rangle\langle p_N|\int\mathrm{d}q_{N-1}|q_{N-1}\rangle\langle q_{N-1}|e^{-\frac{i}{\hbar}\widehat{O}\frac{\alpha}{N}}$$

$$\int\frac{\mathrm{d}p_{N-1}}{2\pi\hbar}|p_{N-1}\rangle\langle p_{N-1}|\cdots\int\mathrm{d}q_1|q_1\rangle\langle q_1|e^{-\frac{i}{\hbar}\widehat{O}\frac{\alpha}{N}}\int\frac{\mathrm{d}p_1}{2\pi\hbar}|p_1\rangle\langle p_1|q_0\rangle.$$

This leads to the following expression

$$\langle q|\exp\left[-\frac{i}{\hbar}\widehat{O}\alpha\right]|q_0\rangle = \prod_{j=1}^{N-1}\int\mathrm{d}q_j\prod_{j=1}^{N}\int\frac{\mathrm{d}p_j}{2\pi\hbar}\exp\left[\frac{i}{\hbar}\mathcal{O}\right],$$

where $\mathcal{O} = \sum_{j=1}^{N}\left[p_j(q_j - q_{j-1}) - \frac{\alpha}{N}O(q_j, p_j)\right]$. The transition amplitude above can formally be written in a path integral form as

$$\langle q|\exp\left[-\frac{i}{\hbar}\widehat{O}\alpha\right]|q_0\rangle = \int_{q_0}^{q}\mathcal{D}''q\mathcal{D}p\exp\frac{i}{\hbar}\int_{0}^{\alpha}\mathrm{d}\bar{\alpha}\left[p\frac{\mathrm{d}q}{\mathrm{d}\bar{\alpha}} - O(q,p)\right].$$

$$(5.109)$$

In this way we have built a *"functional"* expression from the operator $\widehat{O}$. Now we can apply to the functional on the r.h.s. of Eqn.(5.109) the following dequantization rules:

$$\begin{cases} \int\mathrm{d}\bar{\alpha} \longrightarrow i\hbar\int\mathrm{d}\bar{\alpha}\mathrm{d}\theta\mathrm{d}\bar{\theta}, \\ \varphi^a \longrightarrow \Phi^a. \end{cases}$$

What we get is:

$$\int_{Q_0}^{Q}\mathcal{D}''Q\mathcal{D}P\exp i\int_{0}^{\alpha}i\mathrm{d}\bar{\alpha}\mathrm{d}\theta\mathrm{d}\bar{\theta}\left[P\frac{\mathrm{d}Q}{\mathrm{d}\bar{\alpha}} - O(\Phi)\right]. \qquad (5.110)$$

Comparing this with Eqn.(5.85) it is clear that Eqn.(5.110) gives the classical transition amplitude $\langle Q, \alpha|Q_0, 0\rangle$ which in turn can be written as:

$$\langle Q, \alpha|Q_0, 0\rangle = \langle Q|\exp\left[-i\widetilde{\widehat{O}}\alpha\right]|Q_0\rangle,$$

where

$$\widetilde{\widehat{O}} = -i\omega^{ab}\partial_b O\partial_a - i\omega^{ab}\partial_b\partial_d O\bar{c}^d\frac{\partial}{\partial c^a}.$$

This is the classical Lie derivative along the Hamiltonian vector field associated to $O(\varphi^a)$ [Abraham and Marsden (1978)] and is the analog of the $\widehat{\widetilde{\mathcal{H}}}$. So this is the manner to build the *classical* analog of the transformations generated by the quantum operators $O(\widehat{\varphi})$. Again we have applied our usual dequantization rules, but only at the *functional* level.

D) Another object for which we should be careful in naively applying our dequantization rules is the wave function. We had seen in Section 5.4.1 that the substitution $q \to Q$ works at the level of abstract kets $|q\rangle$, in the sense that the states $|Q\rangle$ obtained via our substitution are really the states appearing in the classical transition amplitude (5.85). Now if we try to apply this rule also to the wave functions

$$\langle q|\psi\rangle = \psi(q) \tag{5.111}$$

by substituting on the r.h.s. above $q \to Q$, we would get $\psi(q) \longrightarrow \psi(Q)$. Of course this $\psi(Q)$ does not belong to the "classical" wave functions of KvN, i.e.: $\langle \varphi^a, c^a|\psi\rangle$. $\psi(Q)$ in fact depends on θ and $\bar{\theta}$, while the KvN wave functions are just functions of φ^a and c^a or of q, λ_p, c^q and $\bar{c}_p$, according to the representation we choose. The way out isto realize that actually also the ket $|Q\rangle$ did not depend on θ and $\bar{\theta}$, as explained in Section 5.4.1. In fact $|Q\rangle$ was defined as the eigenstate of the operator $\widehat{Q}$

$$\widehat{Q}(\theta, \bar{\theta})|Q\rangle = Q(\theta, \bar{\theta})|Q\rangle,$$

so the dependence on $\theta, \bar{\theta}$ was contained in the operator $\widehat{Q}$ and in the eigenvalue Q but not in $|Q\rangle$, which, as explained in Section 5.4.1, could be identified with $|q, \lambda_p, c^q, \bar{c}_p\rangle$. This means that the set of wave functions of the operatorial theory lying behind the CPI can be written as

$$\langle Q|\psi\rangle = \langle q, \lambda_p, c^q, \bar{c}_p|\psi\rangle = \psi(q, \lambda_p, c^q, \bar{c}_p).$$

This confirms that the substitution $q \to Q$ cannot be applied to both sides of Eqn.(5.111) since the correct association is given by:

$$\langle q|\psi\rangle = \psi(q)$$

$$\Downarrow$$

$$\langle Q|\psi\rangle = \psi(q, \lambda_p, c^q, \bar{c}_p) \neq \psi(Q).$$

E) The last warning we want to issue on our dequantization rules concerns the extension in the base space $t \longrightarrow (t, \theta, \bar{\theta})$, given by the rule:

$$\int \mathrm{d}t \longrightarrow i\hbar \int \mathrm{d}t\mathrm{d}\theta\mathrm{d}\bar{\theta}.$$

We have seen that this rule works in all the cases we have treated and that even the functional derivative with respect to the current J is strictly related to this rule. The reader may wonder that, if we extend the integration and the whole base space from t to $(t, \theta, \bar{\theta})$ then we should extend also the derivative with respect to t, i.e. d/dt to a sort of combined derivative, like for example the covariant one [West (1986); Deotto and Gozzi (2001)]. This actually does not happen, as it is clear from the weight in Eqn.(5.77), which contains in $L[\Phi]$ the normal time derivative $\int dt\, d\theta\, d\bar{\theta}\, (P\dot{Q} + \cdots)$, or from the Dyson-Schwinger equation (5.103) which also contains the time derivative.

Finally let us summarize our dequantization procedure as follows. In order to pass from QM to CM one must apply the following rules:

1) Given a QM object, in case it is not already written in functional form, build from it a *functional* expression, which either gives the action of that object on the states, or from which the object itself can be derived;

2) Next, in that functional expression perform the following replacements:

   A) the time integration with a proper supertime integration:

   $$\int dt \longrightarrow i\hbar \int dt\, d\theta\, d\bar{\theta},$$

   B) the phase space variables with the superphase space variables:

   $$\varphi^a(t) \longrightarrow \Phi^a(t, \theta, \bar{\theta}),$$

   C) and the functional derivatives with respect to the external currents $J(t)$ with those done with respect to the supercurrents $\mathbb{J}(t, \theta, \bar{\theta})$:

   $$i\hbar \frac{\delta}{\delta J(t)} \longrightarrow \frac{\delta}{\delta \mathbb{J}(t, \theta, \bar{\theta})}.$$

5.5 Superposition in classical mechanics

So far we have considered the path integral formalism as a common framework for classical mechanics and quantum mechanics. As we have seen any quantum system can be described either by a path integral or via its operatorial counterpart. In the operatorial version of QM a system is represented by some physical state belonging to a Hilbert space and for the physical states the superposition principle holds which says that if $|\varphi_A\rangle$ is a state describing the system A and $|\psi_A\rangle$ another state describing the same system A then the state $|\widetilde{\psi}_A\rangle = c_1|\varphi_A\rangle + c_2|\psi_A\rangle$ is also an admissible state for the system A. The reader may get confused by thinking that something similar happens also in the operatorial counterpart of the CPI. In CM we know that a system at some time t is represented by its position in phase space $\varphi^a(t)$ and it does not make sense to think of superposition. If we consider the Hilbert space associated to the CPI we can label an element of this space via the phase space coordinates $\varphi^a(t)$ as well as the ghosts $c^a(t)$, i.e.: $|\psi\rangle = |\varphi^a, c^a\rangle$. The reader thus may wonder what is the meaning of superposing two such states and ask if this results in a new physical state. It will turn out that in CM superposition is prevented by a superselection rule which we shall describe in the following.

There are some limitations on the superposition principles even in QM that may eventually arise from dynamical or kinetical requirements of the theory. For example it is not possible to superimpose two states which are labelled by two different values of the electric charge, neither we can superimpose two states having one half-integer and the other integer spin. This feature was first discovered by Wick, Wightman and Wigner in [Wick *et al.* (1952)]. For more details we refer the reader to [Roman (1965)], [Fonda and Ghirardi (1970)]. As a simple example of this mechanism we can consider the superposition of a particle with spin 1/2 and one with spin 1:

$$|\psi\rangle = |1/2\rangle + |1\rangle. \qquad (5.112)$$

Now suppose we do a rotation of 2π on the system, we have:

$$|\widetilde{\psi}\rangle = -|1/2\rangle + |1\rangle. \qquad (5.113)$$

The vectors $|\psi\rangle$ and $|\widetilde{\psi}\rangle$ represent the same physical state since we have just performed a rotation of 2π on the system. Nevertheless they are two different vectors in the Hilbert space. Therefore we have lost the one-to-one correspondence between vectors of the Hilbert space and the physical states. In order to avoid this situation a superselection rule has to be considered. In general the superselection rules are produced by the following mechanism [Roman (1965)]: if there exists an operator A different from the identity and which commutes with all the observables of a theory, then the total Hilbert space $\mathcal{H}$ of the theory is naturally decomposed in the direct sum of

Hilbert spaces $\mathcal{H}_{a_i}$ given by the eigenvarieties associated to each eigenvalue a_i of the operator A, called the superselection operator:

$$\mathcal{H} = \mathcal{H}_{a_1} \oplus \cdots \oplus \mathcal{H}_{a_n} .$$

The most important fact [Roman (1965)] is that the allowed or "physical" states cannot be linear superpositions of states $|\phi_{a_i}\rangle$ belonging to different Hilbert spaces $\mathcal{H}_{a_i}$:

$$|\psi_{phys}\rangle \neq \sum_i c_i |\phi_{a_i}\rangle .$$

For more details on why this happens we refer the reader to [Roman (1965)].

Such superselection mechanism is crucial in order to understand the nature of the superposition in the CPI. One of the KvN postulates states that the observables of CM are the functions of only the phase space variables $\widehat{\varphi}$, i.e.: $\widehat{O}(\widehat{\varphi})$. This postulate is different from the quantum mechanical one which states that the observables have the only feature of being Hermitian operators. If, after having built a scalar product with a proper Hermicity concept [Deotto *et al.* (2003a,b)], we apply the same postulates of QM to CM then we could say that functions depending not only on $\widehat{\varphi}$ but also on $\widehat{\lambda}$, $\widehat{c}$, $\widehat{\bar{c}}$ could be considered observables in the CPI provided that they are Hermtian:

$$O^\dagger(\widehat{\varphi}, \widehat{\lambda}, \widehat{c}, \widehat{\bar{c}}) = O(\widehat{\varphi}, \widehat{\lambda}, \widehat{c}, \widehat{\bar{c}}). \tag{5.114}$$

Anyhow this would lead to problems for CM. In fact, even neglecting c, $\bar{c}$ and restricting to observables of the form $\widehat{O}^\dagger(\varphi, \lambda) = \widehat{O}(\varphi, \lambda)$, we notice that many of them would be non-commuting because $\widehat{\varphi}$ and $\widehat{\lambda}$ do not commute (see Eqn.(5.32)). This non-commuting feature would lead to interference effects which have never been detected in CM. This is the reason that induced KvN to restrict the observables to just those depending on $\widehat{\varphi}$, i.e.: $\widehat{O}(\widehat{\varphi})$. In this section we want to see if we really need this postulate which restricts *by hand* the observables or if we can resort to something deeper. In doing so we will discover also the superselection mechanism which forbids the superposition in CM.

We have seen in the previous sections that the CPI has a lot of *global* symmetries whose charges were given in Eqn.(5.40), but the CPI can be written in terms of superfields like:

$$Z = \int \mathcal{D}\Phi \exp\left[i \int_{t_0}^{t} id\tau d\theta d\bar{\theta}\, L[\Phi]\right] \tag{5.115}$$

and this expression leads to new symmetries. For example let us consider the following transformation:

$$\begin{cases} \varphi^a \longrightarrow \varphi^a + \varepsilon(t)\theta c^a \\ c^a \longrightarrow c^a - \varepsilon(t)c^a \end{cases} \tag{5.116}$$

where $\varepsilon(t)$ is an infinitesimal parameter depending on t and θ is the Grassmannian partner of time used in the superfield (5.59). They are local (or gauge) symmetries because they depend on the base space variables t, θ, $\bar{\theta}$. It is easy to check that the superfield (5.59) remains invariant under the trasformation (5.116) and the same for its time derivative. There are many other similar invariances like these and we refer to [Cattaruzza and Gozzi (2012)] for a complete discussion on this point. Here we will present only two more besides those of Eqn.(5.116). They are:

$$\begin{cases} \varphi^a \longrightarrow \varphi^a + \varepsilon(t)\bar{\theta}\omega^{ab}\bar{c}_b \\ \bar{c}_b \longrightarrow \bar{c}_b - \varepsilon(t)\bar{c}_b \end{cases} \tag{5.117}$$

and

$$\begin{cases} \varphi^a \longrightarrow \varphi^a + i\varepsilon(t)\bar{\theta}\theta\varphi^a \\ \lambda_b \longrightarrow \lambda_b - \varepsilon(t)\omega_{bc}\varphi^c. \end{cases} \tag{5.118}$$

It is easy to prove that the only object invariant under all these three trasformations is the superfield and its time derivative. Actually the first transformation (5.116) leaves invariant also the sub-piece $\varphi^a + \theta c^a$ but this is not invariant under the trasformations (5.117) and (5.118). Similarly (5.117) leaves invariant, besides the superfield, also the sub-piece $\varphi^a + \bar{\theta}\omega^{ab}\bar{c}_b$, but this is not left invariant by the trasformations (5.116) and (5.118). Similarly for (5.118). So the only object that is simultaneously invariant under the trasformations (5.116), (5.117) and (5.118) is the superfield and its time derivative. As a consequence the action in Eqn.(5.115) is also invariant and so we can say that the trasformatons (5.116), (5.117), (5.118) are gauge symmetries of our system.

Let us now turn to the observables. As there are gauge symmetries in our system, the acceptable observables are only those invariant under the same gauge symmetries. As the superfield and its time derivative are the only objects invariant under (5.116), (5.117), (5.118) we have that the only acceptable observables are:

$$\widetilde{O}(\widehat{\Phi}^a, \dot{\widehat{\Phi}}^a) \tag{5.119}$$

Using the equation of motion for Φ^a, i.e.: $\dot{\Phi}^a = \omega^{ab}\partial_b H[\Phi]$ we can replace $\dot{\Phi}$ by a function of Φ in $\widetilde{O}$, and so we can conclude that the acceptable observables are of the form:

$$O(\widehat{\Phi}^a) \tag{5.120}$$

with $O \neq \tilde{O}$. At this point the reader may point out that the $O(\widehat{\Phi}^a)$ are not the $O(\widehat{\varphi}^a)$ which were the observables we wanted to get. Actually the functions (5.120) are observables which depend explicitly on the two Grassmannian partners of time θ and $\bar{\theta}$. So with respect to these extra times we can say that the observables (5.120) are in the "Heisenberg" picture. As we have seen in Section 5.4 the "Hamiltonians" associated to the extra time variables θ and $\bar{\theta}$ are the Q_{BRS} and $\bar{Q}_{BRS}$ because they generate a translation in θ and $\bar{\theta}$ respectively. So the transformation from the Heisenberg picture to the Schrödinger one in θ and $\bar{\theta}$ is:

$$e^{-\theta Q_{BRS} - \bar{Q}_{BRS}\bar{\theta}} O_H(\widehat{\Phi}^a) e^{\theta Q_{BRS} + \bar{Q}_{BRS}\bar{\theta}} \equiv O_S = O(\widehat{\varphi}). \qquad (5.121)$$

This is the most important step of the theorem we want to prove. Basically Eqn.(5.121) says that the acceptable observables are the $O(\widehat{\Phi}^a)$ and these are none other than the Heisenberg picture in θ and $\bar{\theta}$ of the standard observables of CM: $O(\widehat{\varphi}^a)$. So we can conclude that the local symmetries (5.116), (5.117), (5.118) are eventually responsible for the commuting character of CM because all the $O(\widehat{\varphi}^a)$ commute among themselves. So we can remove the 5th postulate of KvN. The reader may wonder how that 5th postulate could suppress superposition. The reason is the following: it is crucial to note that $\widehat{\varphi}^a$ commutes with all the observables $O(\widehat{\varphi})$ so it is a *superselection* operator [Roman (1965)]. This triggers the *superselection* mechanism which says that the physical Hilbert space of the system is given by an eigenvariety of the superselection operator, i.e.: $\widehat{\varphi}^a$. In our case we have:

$$\widehat{\varphi}^a |\varphi_0^a\rangle = \varphi_0^a |\varphi_0^a\rangle \qquad (5.122)$$

where φ_0^a is a particular point in the phase space. This eigenvariety in the $|\varphi\rangle$ basis is basically a Dirac delta state:

$$\langle \varphi^a |\varphi_0^a\rangle = \delta(\varphi^a - \varphi_0^a). \qquad (5.123)$$

Another eigenvariety is given by:

$$\widehat{\varphi}^a |\varphi_1^a\rangle = \varphi_1^a |\varphi_1^a\rangle \qquad (5.124)$$

where φ_1^a is a particular point in the phase space. Of course, as Eqns.(5.122) and (5.124) gives different eigenvarieties, according to the superselection principle we cannot do linear superposition of states belonging to different eigenvarieties, i.e.: the state

$$|\widetilde{\varphi^a}\rangle \equiv |\varphi_0^a\rangle + |\varphi_1^a\rangle \qquad (5.125)$$

is not a *physical* one. This is the basic reason why in CM there is no superposition and as a consequence no interference. This analysis is limited to the 0-form sector of the theory, i.e.: $c = \bar{c} = 0$, but the same reasoning applies also when c and $\bar{c}$ are present and details can be found in [Gozzi and Pagani (2010)].

The main conclusion we like to draw is that the non-superposition and non-interference in CM are basically due to some local universal invariances (5.116), (5.117), (5.118) present in CM. Let us point out that in QM we do not have anymore those invariances because, as we have seen in Section 5.4, QM is obtained from CM by a dimensional reduction which sends $(\theta, \bar{\theta}) \longrightarrow 0$. If these variables, θ, $\bar{\theta}$, are zero then the local symmetries disappear and the transformations (5.116), (5.117), (5.118) are reduced to the identity.

Appendix A

Asynchronous variation of the action

Let us consider a system described by the Lagrangian $L\left(\mathbf{q}\left(t\right),\dot{\mathbf{q}}\left(t\right),t\right)$. Adopting the usual notation we recall that the *action functional* is defined as

$$S\left[\mathbf{q}\left(t\right)\right] \equiv \int_{t_1}^{t_2} \mathrm{d}t\, L\left(\mathbf{q}\left(t\right),\dot{\mathbf{q}}\left(t\right),t\right). \qquad (\text{A.1})$$

We are interested in the solution of the following variational problem

$$\delta S\left[\mathbf{q}\left(t\right)\right] = 0, \qquad (\text{A.2})$$

where we perform a *synchronous* variation and the trajectories have the same fixed endpoints, i.e.:

$$\left\{ \mathbf{q}\left(t\right) : \mathbf{q}\left(t\right) \in C^k\left(\mathbb{R}\right)\ \left(k \geq 2\right),\ \mathbf{q}\left(t_i\right) = \mathbf{q}_i\ \left(i = 1,2\right) \right\}. \qquad (\text{A.3})$$

We call $\mathbf{q}_{cl}\left(t\right)$ such a solution, meaning that $\delta S\left[\mathbf{q}_{cl}\left(t\right)\right] = 0$. We refer to $\mathbf{q}_{cl}\left(t\right)$ as the classical trajectory. The Hamilton's principle tells us that $\mathbf{q}_{cl}\left(t\right)$ satisfies the Euler-Lagrange equations with the following boundary conditions $\mathbf{q}\left(t_i\right) = \mathbf{q}_i\ \left(i = 1,2\right)$.

Given any $\mathbf{q}\left(t\right)$ we construct a synchronously varied trajectory $\mathbf{q}_{syn}\left(t\right)$ defined in the following manner:

$$\mathbf{q}_{syn}\left(t\right) \equiv \mathbf{q}\left(t\right) + \varepsilon\eta\left(t\right) \equiv \mathbf{q}\left(t\right) + \delta\mathbf{q}\left(t\right), \qquad (\text{A.4})$$

where $\eta\left(t_i\right) = 0\ \left(i = 1,2\right)$ and $\varepsilon \neq 0$ but it is infinitesimal (see Fig. A.1). We call $\delta\mathbf{q}\left(t\right)$ the *variation at equal times* or synchronous variation.

The synchronous variation of the action functional, also called *variation-δ*, is defined as

$$\delta S\left[\mathbf{q}\left(t\right)\right] \equiv S\left[\mathbf{q}\left(t\right) + \delta\mathbf{q}\left(t\right)\right] - S\left[\mathbf{q}\left(t\right)\right], \qquad (\text{A.5})$$

where $\delta\mathbf{q}\left(t\right)$ is infinitesimal.

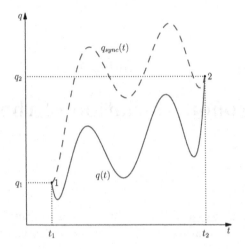

Fig. A.1   Synchronously varied motion.

Now we can construct a more general class of variations and consider trajectories which have endpoints with times and coordinates different from t_1, t_2, $\mathbf{q}_1$ and $\mathbf{q}_2$. Given a $\mathbf{q}(t)$ we consider the trajectory

$$\mathbf{q}'(t) \equiv \mathbf{q}(t) + \varepsilon \eta(t), \qquad (A.6)$$

where now $\eta(t)$ is not subject to any condition at the endpoints. We define the *variation at different times* as

$$\Delta \mathbf{q}(t) \equiv \mathbf{q}'(t') - \mathbf{q}(t), \qquad (A.7)$$

where with $\mathbf{q}'(t')$ we denote the varied trajectory (A.6) parametrized by a new time t'. If we suppose that $\Delta t \equiv t' - t$ and ε are infinitesimal then, at first order, we find

$$\Delta \mathbf{q}(t) = [\mathbf{q}'(t') - \mathbf{q}'(t)] + [\mathbf{q}'(t) - \mathbf{q}(t)]$$
$$= \dot{\mathbf{q}}'(t)\,\Delta t + \delta \mathbf{q}(t) = \dot{\mathbf{q}}(t)\,\Delta t + \delta \mathbf{q}(t). \qquad (A.8)$$

The *asynchronous variation* of the trajectory is therefore defined as

$$\mathbf{q}_{asyn}(t) \equiv \mathbf{q}(t) + \Delta \mathbf{q}(t). \qquad (A.9)$$

In order to understand better this point see Fig. A.2. Let us consider the curve $\mathbf{q}(t)$, which satisfies $\mathbf{q}(t_i) = \mathbf{q}_i$ $(i = 1, 2)$, and another curve, which we call $\mathbf{q}_{asyn}(t)$, which is completely arbitrary (provided that it is infinitesimally close to $\mathbf{q}(t)$ at any point). The variation provides a one-to-one correspondence between the points belonging to $\mathbf{q}(t)$ and those belonging to $q_{asyn}(t')$. Such correspondence is naturally induced by the temporal order parametrizing the two curves. Thus Δq at a generic point P is simply given by

$$\Delta \mathbf{q}(t)|_P = \mathbf{q}_{asyn}(t')|_P - \mathbf{q}(t)|_P. \qquad (A.10)$$

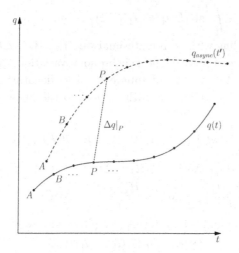

Fig. A.2 Asynchronously varied motion.

In analogy to what we have seen in Eqn.(A.5) we define the asynchronous variation of the action functional, also called *variation-Δ*, as

$$\Delta S\left[\mathbf{q}(t)\right] \equiv S\left[\mathbf{q}(t) + \Delta \mathbf{q}(t)\right] - S\left[\mathbf{q}(t)\right], \qquad (A.11)$$

where $\Delta \mathbf{q}(t)$ is infinitesimal. Let us compute explicitly such variation:

$$\Delta S\left[\mathbf{q}(t)\right] = S\left[\mathbf{q}_{asyn}(t)\right] - S\left[\mathbf{q}(t)\right] = S\left[\mathbf{q}(t) + \Delta \mathbf{q}(t)\right] - S\left[\mathbf{q}(t)\right]$$

$$= \int_{t_1'}^{t_2'} \mathrm{d}t\, L\left(\mathbf{q}(t) + \Delta \mathbf{q}(t), \dot{\mathbf{q}}(t) + \Delta \dot{\mathbf{q}}(t), t\right)$$

$$- \int_{t_1}^{t_2} \mathrm{d}t\, L\left(\mathbf{q}(t), \dot{\mathbf{q}}(t), t\right), \qquad (A.12)$$

where the t_i' denotes the extremities of the trajectory $q_{asyn}(t)$, which have been varied via an asynchronous variation with respect to the unprimed endpoints of $\mathbf{q}\left(t\right)$. Now we rewrite the variation (A.12) as follows

$$
\Delta S\left[\mathbf{q}\left(t\right)\right] = \left[\int_{t_1'}^{t_2'} dt\, L\left(\mathbf{q}\left(t\right)+\Delta\mathbf{q}\left(t\right),\dot{\mathbf{q}}\left(t\right)+\Delta\dot{\mathbf{q}}\left(t\right),t\right)\right.
$$

$$
\left.- \int_{t_1'}^{t_2'} dt\, L\left(\mathbf{q}\left(t\right),\dot{\mathbf{q}}\left(t\right),t\right)\right] + \left[\int_{t_1'}^{t_2'} dt\, L\left(\mathbf{q}\left(t\right),\dot{\mathbf{q}}\left(t\right),t\right)\right.
$$

$$
\left.- \int_{t_1}^{t_2} dt\, L\left(\mathbf{q}\left(t\right),\dot{\mathbf{q}}\left(t\right),t\right)\right]. \tag{A.13}
$$

Since we are dealing with an infinitesimal variation (i.e.: Δt is infinitesimal at first order) we recognize the synchronous variation (A.5) in the first square bracket. In the second square bracket, denoting with $F_L\left(t\right)$ the primitive of the Lagrangian L with respect to the integration over t, we have

$$
\int_{t_1+\Delta t_1}^{t_2+\Delta t_2} dt\, L\left(\mathbf{q}\left(t\right),\dot{\mathbf{q}}\left(t\right),t\right) - \int_{t_1}^{t_2} dt\, L\left(\mathbf{q}\left(t\right),\dot{\mathbf{q}}\left(t\right),t\right)
$$

$$
= F_L\left(t_2+\Delta t_2\right) - F_L\left(t_1+\Delta t_1\right) - F_L\left(t_2\right) + F_L\left(t_1\right)
$$

$$
\approx \left.\frac{dF_L\left(t\right)}{dt}\right|_{t=t_2}\Delta t_2 - \left.\frac{dF_L\left(t\right)}{dt}\right|_{t=t_1}\Delta t_1. \tag{A.14}
$$

Therefore, up to infinitesimal terms of higher order, we have

$$
\Delta S\left[\mathbf{q}\left(t\right)\right] = \delta S\left[\mathbf{q}\left(t\right)\right] + \left[L\left(\mathbf{q}\left(t\right),\dot{\mathbf{q}}\left(t\right),t\right)\Delta t\right]_{t=t_1}^{t=t_2}. \tag{A.15}
$$

If we now use the equations of motion (A.15) for the synchronous variation we find that

$$
\Delta S\left[\mathbf{q}\left(t\right)\right] = \int_{t_1}^{t_2} dt\, \delta\mathbf{q}\left(t\right)\cdot\left[\frac{\partial L}{\partial\mathbf{q}} - \frac{d}{dt}\left(\frac{\partial L}{\partial\dot{\mathbf{q}}}\right)\right]
$$

$$
+ \left[L\left(\mathbf{q}\left(t\right),\dot{\mathbf{q}}\left(t\right),t\right)\Delta t + \frac{\partial L}{\partial\dot{\mathbf{q}}}\cdot\delta\mathbf{q}\left(t\right)\right]_{t=t_1}^{t=t_2}. \tag{A.16}
$$

For the ease of notation let us denote

$$
C \equiv L\left(\mathbf{q}\left(t\right),\dot{\mathbf{q}}\left(t\right),t\right)\Delta t + \frac{\partial L}{\partial\dot{\mathbf{q}}}\cdot\delta\mathbf{q}\left(t\right). \tag{A.17}
$$

Using Eqn.(A.8) and the definition of Hamiltonian we have that

$$
C = L\left(\mathbf{q}\left(t\right),\dot{\mathbf{q}}\left(t\right),t\right)\Delta t + \frac{\partial L}{\partial\dot{\mathbf{q}}}\cdot\left[\Delta\mathbf{q}\left(t\right)-\dot{\mathbf{q}}\left(t\right)\Delta t\right]
$$

$$
= \left[\dot{\mathbf{q}}\left(t\right)\cdot\frac{\partial L}{\partial\dot{\mathbf{q}}} - H\right]\Delta t + \frac{\partial L}{\partial\dot{\mathbf{q}}}\cdot\left[\Delta\mathbf{q}\left(t\right)-\dot{\mathbf{q}}\left(t\right)\Delta t\right]
$$

$$
= \frac{\partial L}{\partial\dot{\mathbf{q}}}\cdot\Delta\mathbf{q}\left(t\right) - H\left(\mathbf{q}\left(t\right),\mathbf{p}\left(t\right),t\right)\Delta t
$$

$$
= \mathbf{p}\left(t\right)\cdot\Delta\mathbf{q}\left(t\right) - H\left(\mathbf{q}\left(t\right),\mathbf{p}\left(t\right),t\right)\Delta t. \tag{A.18}
$$

Thus finally

$$\Delta S\left[\mathbf{q}\left(t\right)\right] = \int_{t_1}^{t_2} dt\; \delta \mathbf{q}\left(t\right) \cdot \left[\frac{\partial L}{\partial \mathbf{q}} - \frac{d}{dt}\left(\frac{\partial L}{\partial \dot{\mathbf{q}}}\right)\right]$$
$$+ \left[\mathbf{p}\left(t\right)\cdot \Delta \mathbf{q}\left(t\right) - H\left(\mathbf{q}\left(t\right),\mathbf{p}\left(t\right),t\right)\Delta t\right]_{t=t_1}^{t=t_2}. \quad (A.19)$$

Let us study the action from another point of view. We consider the action as a quantity which characterizes the motion along the classical trajectories emanating from a common initial point but having the final extremity not fixed. Thus we define the *action function* as

$$S_{cl}\left(\mathbf{q}_{cl}\left(t\right),t\right) \equiv \int_{t_1}^{t} dt'\; L\left(\mathbf{q}_{cl}\left(t'\right),\dot{\mathbf{q}}_{cl}\left(t'\right),t'\right), \quad (A.20)$$

where the integral is computed along a trajectory $\mathbf{q}_{cl}\left(t\right)$, which is the solution of the canonical equations of motion with initial condition

$$\mathbf{q}_{cl}\left(t_1\right) = \mathbf{Q}, \quad (A.21a)$$
$$\mathbf{p}_{cl}\left(t_1\right) = \mathbf{P}. \quad (A.21b)$$

Changing the boundary condition (A.21b) we obtain trajectories having different final points (for $t = t_2$). Such trajectories are connected to particular asynchronous variations, for which $\Delta \mathbf{q}_{cl}\left(t_1\right) = 0$ and $\Delta t_1 = 0$. Since we are working with the function (A.20), and not with the functional (A.1), on classical trajectories the equation (A.19) becomes

$$dS_{cl}\left(\mathbf{q}_{cl}\left(t\right),t\right) = \mathbf{p}_{cl}\left(t\right)\cdot \Delta \mathbf{q}_{cl}\left(t\right) - H\left(\mathbf{q}_{cl}\left(t\right),\mathbf{p}_{cl}\left(t\right),t\right)\Delta t. \quad (A.22)$$

Thus we can conclude that

$$\mathbf{p}_{cl}\left(t\right) = \frac{\partial S_{cl}}{\partial \mathbf{q}_{cl}\left(t\right)} \quad \text{and} \quad H\left(\mathbf{q}_{cl}\left(t\right),\mathbf{p}_{cl}\left(t\right),t\right) = -\frac{\partial S_{cl}}{\partial t}, \quad (A.23)$$

which can be rewritten as a unique partial differential equation

$$H\left(\mathbf{q}_{cl}\left(t\right),\frac{\partial S_{cl}}{\partial \mathbf{q}_{cl}\left(t\right)},t\right) + \frac{\partial S_{cl}}{\partial t} = 0 \quad (A.24)$$

which is the *Hamilton-Jacobi equation*. This is the same equation that holds for the generating function $F\left(\mathbf{q},\mathbf{P},t\right)$ in Eqn.(3.19)[1]. Therefore we have shown that the action function is the generating function of the canonical transformation

$$\left(\mathbf{q},\mathbf{p}\right) \longrightarrow \left(\mathbf{Q},\mathbf{P}\right) \quad (A.25)$$

which gives a vanishing Hamiltonian in the transformed variables.

---

[1] Here we use the subscript cl to stress the fact that the trajectory $\mathbf{q}\left(t\right)$ is well defined.

Finally we observe that

$$\mathbf{q}_{cl}\left(t\right) = \mathbf{q}_{cl}\left(\mathbf{Q},\mathbf{P},t\right), \tag{A.26}$$

hence, for fixed $\mathbf{Q}$, $S_{cl} = S_{cl}\left(\mathbf{q}_{cl},\mathbf{P},t\right)$ depends on the same variables as F. However it may be convenient to express S_{cl} as a function of other variables, for instance $S_{cl} = S_{cl}\left(\mathbf{q}_{cl},\mathbf{Q},t\right)$. Let us suppose that we know the explicit form of the canonical transformation (A.25). We compute the differential dS_{cl} using the relations (3.18a), (3.18b), (3.18c) and recalling that $K = 0$ (see Section 3.1.2) we get

$$dS_{cl}\left(\mathbf{q}_{cl},\mathbf{P},t\right) = \frac{\partial S_{cl}}{\partial \mathbf{q}_{cl}} \cdot d\mathbf{q}_{cl} + \frac{\partial S_{cl}}{\partial \mathbf{P}} \cdot d\mathbf{P} + \frac{\partial S_{cl}}{\partial t} dt$$

$$= \mathbf{p}_{cl} \cdot d\mathbf{q}_{cl} + \mathbf{Q} \cdot d\mathbf{P} - H dt. \tag{A.27}$$

We perform now a Legendre transform

$$S_{cl}\left(\mathbf{q}_{cl},\mathbf{Q},t\right) \equiv S_{cl}\left(\mathbf{q}_{cl},\mathbf{P},t\right) - \mathbf{Q}\cdot\mathbf{P} \tag{A.28}$$

and we find that

$$dS_{cl}\left(\mathbf{q}_{cl},\mathbf{Q},t\right) = dS_{cl}\left(\mathbf{q}_{cl},\mathbf{P},t\right) - \mathbf{Q}\cdot d\mathbf{P} - \mathbf{P}\cdot d\mathbf{Q}$$

$$= \mathbf{p}_{cl} \cdot d\mathbf{q}_{cl} - \mathbf{P}\cdot d\mathbf{Q} - H dt. \tag{A.29}$$

Hence we obtain the relation for the initial momentum

$$\mathbf{P} = -\frac{\partial S_{cl}\left(\mathbf{q}_{cl},\mathbf{Q},t\right)}{\partial \mathbf{Q}}. \tag{A.30}$$

The equation for the function $f(t_2, t_1)$ introduced in Section 2.6

We want to derive the Cauchy problem for the function $f(t_2, t_1)$, introduced in Section 2.6 and Section 3.3.2. We recall that

$$f(t_2, t_1) = \lim_{N \to \infty} \varepsilon \left(\frac{2\hbar i \varepsilon}{m} \right)^N \det \underline{\underline{\sigma}} \tag{B.1}$$

and the definition of $\underline{\underline{\sigma}}$ is

$$\underline{\underline{\sigma}} \equiv \frac{i}{\hbar} \left[\frac{m}{2} \frac{1}{\varepsilon} \begin{pmatrix} -2 & 1 & & 0 \\ 1 & -2 & 1 & \\ & & \ddots & 1 \\ 0 & & 1 & -2 \end{pmatrix} - \varepsilon \begin{pmatrix} c(t_{\alpha_1}) & & & 0 \\ & c(t_{\alpha_2}) & & \\ & & \ddots & \\ 0 & & & c(t_{\alpha_N}) \end{pmatrix} \right],$$

where $c(t_{\alpha_i})$ are the coefficients appearing in the matrix $\underline{\underline{B}}$ of Eqn.(3.94). We find that

$$\left(\frac{2\hbar i \varepsilon}{m} \right)^N \det \underline{\underline{\sigma}} = \det \left[\begin{pmatrix} 2 & -1 & & 0 \\ -1 & 2 & -1 & \\ & \ddots & \ddots & \ddots \\ & & -1 & 2 & -1 \\ 0 & & & -1 & 2 \end{pmatrix} \right.$$

$$\left. + \frac{2}{m} \varepsilon^2 \begin{pmatrix} c(t_{\alpha_1}) & & & & 0 \\ & c(t_{\alpha_2}) & & & \\ & & \ddots & & \\ & & & c(t_{\alpha_{N-1}}) & \\ 0 & & & & c(t_{\alpha_N}) \end{pmatrix} \right] \equiv \det \underline{\underline{\tilde{\sigma}}} \equiv d_N, \tag{B.2}$$

where d_N denotes the determinant of the matrix in squared brackets $\underline{\underline{\tilde{\sigma}}}$. Thus let us consider the minors of $\underline{\underline{\tilde{\sigma}}}$, and let d_j be the determinant of the $j \times j$ matrix built out of the first j rows and columns of $\underset{=}{\tilde{\sigma}}$. It is straightforward to see that the following recursive formula holds

$$d_{j+1} = \left[2 + \frac{2\varepsilon^2}{m} c\left(t_{\alpha_{j+1}}\right) \right] d_j - d_{j-1} \quad (j = 1, ..., N), \qquad (B.3)$$

with $d_0 = 1$. It is easy to convince ourselves of Eqn.(B.3) applying the Laplace's formula for the determinant of $\underset{=}{\tilde{\sigma}}$, whose explicit expression is

$$\underset{=}{\tilde{\sigma}} = \begin{pmatrix} 2 + \frac{2}{m}\varepsilon^2 c\left(t_{\alpha_1}\right) & -1 & & & 0 \\ -1 & 2 + \frac{2}{m}\varepsilon^2 c\left(t_{\alpha_2}\right) & -1 & & \\ & \ddots & \ddots & \ddots & \\ & & & & -1 \\ 0 & & & -1 & 2 + \frac{2}{m}\varepsilon^2 c\left(t_{\alpha_N}\right) \end{pmatrix}. \qquad (B.4)$$

Equation (B.3) can be rewritten as

$$m \frac{d_{j+1} - 2d_j + d_{j-1}}{\varepsilon^2} = 2c\left(t_{\alpha_{j+1}}\right) d_j. \qquad (B.5)$$

Setting $\varphi\left(t_1 + j\varepsilon\right) \equiv \varepsilon d_j$, in the limit $N \to \infty$ and $\varepsilon \to 0$ we find that φ has to satisfy the following differential equation[1]

$$m\ddot{\varphi}\left(t\right) - 2c\left(t\right)\varphi\left(t\right) = 0. \qquad (B.6)$$

For what concerns the boundary conditions we obtain

$$\varphi\left(t_1\right) = \lim_{\varepsilon \to 0} \varepsilon d_0 = 0 \qquad (B.7a)$$

$$\dot{\varphi}\left(t_1\right) = \lim_{\varepsilon \to 0} \varepsilon \left(\frac{d_1 - d_0}{\varepsilon}\right) = \lim_{\varepsilon \to 0} \left[2 + \frac{2\varepsilon^2}{m}c\left(t_{\alpha_1}\right) - 1\right] = 1 \quad (B.7b)$$

Finally we observe that from Eqn.(B.1) we have that

$$f\left(t_2, t_1\right) = \lim_{N \to \infty} d_N = \varphi\left(t_2\right). \qquad (B.8)$$

Moreover we note that

$$c\left(t\right) = -\frac{1}{2}\frac{\partial^2 V\left(x, t\right)}{\partial x^2}. \qquad (B.9)$$

Therefore the Cauchy problem for the function $f\left(t_2, t_1\right)$ is given by

$$\begin{cases} \dfrac{\partial^2 f\left(t_2, t_1\right)}{\partial t_2^2} + \dfrac{1}{m}\dfrac{\partial^2 V\left(x, t_2\right)}{\partial x^2}f\left(t_2, t_1\right) = 0 \\[4mm] f\left(t_1, t_1\right) = 0 \text{ and } \left.\dfrac{\partial f\left(t_2, t_1\right)}{\partial t_2}\right|_{t_2 = t_1} = 1. \end{cases} \qquad (B.10)$$

---

[1]To pass from $t_1 + j\varepsilon$ to the continuous variable t we have used the definition $\varepsilon = \left(t_2 - t_1\right)/\left(N + 1\right)$.

Appendix C

Variational calculus in the discrete formalism

Let us consider the discretized form of the action S_N given by Eqn.(3.81). We want to deduce the Euler-Lagrange equations and show that $\delta S_N[x_{cl}] = 0$. We limit ourselves to consider a Lagrangian of the form

$$L = \frac{m}{2}\dot{x}^2 - V(x,t).\tag{C.1}$$

Let $\delta x = \zeta$ be the infinitesimal variation of the trajectory, then by definition the first variation of S_N is given by

$$\delta S_N[x] = S_N[x + \zeta] - S_N[x].\tag{C.2}$$

Explicitly we find that

$$\delta S_N[x] = \varepsilon \sum_{k=1}^{N+1} \left[\frac{m}{2} \left(\frac{x_{\alpha_k} + \zeta_{\alpha_k} - x_{\alpha_{k-1}} - \zeta_{\alpha_{k-1}}}{\varepsilon} \right)^2 - V(x_{\alpha_k} + \zeta_{\alpha_k}, t_{\alpha_k}) \right]$$
$$-\varepsilon \sum_{k=1}^{N+1} \left[\frac{m}{2} \left(\frac{x_{\alpha_k} - x_{\alpha_{k-1}}}{\varepsilon} \right)^2 - V(x_{\alpha_k}, t_{\alpha_k}) \right],\tag{C.3}$$

and expanding the square in the kinetic terms[1]

$$\delta S_N[x] = \varepsilon \sum_{k=1}^{N+1} \left[\frac{m}{2\varepsilon^2} \left(x_{\alpha_k}^2 + \zeta_{\alpha_k}^2 + x_{\alpha_{k-1}}^2 + \zeta_{\alpha_{k-1}}^2 + 2x_{\alpha_k}\zeta_{\alpha_k} - 2x_{\alpha_k}x_{\alpha_{k-1}} \right.\right.$$
$$-2x_{\alpha_k}\zeta_{\alpha_{k-1}} - 2x_{\alpha_{k-1}}\zeta_{\alpha_k} - 2\zeta_{\alpha_k}\zeta_{\alpha_{k-1}} + 2x_{\alpha_{k-1}}\zeta_{\alpha_{k-1}} - x_{\alpha_k}^2$$
$$\left.\left. -x_{\alpha_{k-1}}^2 + 2x_{\alpha_k}x_{\alpha_{k-1}} \right) - \frac{\partial V}{\partial x_{\alpha_k}}\zeta_{\alpha_k} \right]\tag{C.4}$$

---

[1] In all the formulas which follow V is evaluated at time t_{α_k} unless otherwise stated. Moreover it is understood that:

$$\frac{\partial V}{\partial x_{\alpha_k}} = \frac{\partial V(x)}{\partial x}\bigg|_{x=x_{\alpha_k}}.$$

and neglecting terms of higher order we have that

$$\delta S_N[x] = \varepsilon \sum_{k=1}^{N+1} \left[\frac{m}{\varepsilon^2} \left(x_{\alpha_k} \zeta_{\alpha_k} - x_{\alpha_k} \zeta_{\alpha_{k-1}} - x_{\alpha_{k-1}} \zeta_{\alpha_k} + x_{\alpha_{k-1}} \zeta_{\alpha_{k-1}} \right) \right.$$
$$\left. - \frac{\partial V}{\partial x_{\alpha_k}} \zeta_{\alpha_k} \right]$$
$$= \varepsilon \sum_{k=1}^{N+1} \left[\frac{m}{\varepsilon^2} \left(\zeta_{\alpha_k} - \zeta_{\alpha_{k-1}} \right) \left(x_{\alpha_k} - x_{\alpha_{k-1}} \right) - \frac{\partial V}{\partial x_{\alpha_k}} \zeta_{\alpha_k} \right]. \quad \text{(C.5)}$$

This can be rewritten as

$$\delta S_N[x] = \varepsilon \sum_{k=1}^{N+1} \left[m \left(\frac{\zeta_{\alpha_k} - \zeta_{\alpha_{k-1}}}{\varepsilon} \right) \left(\frac{x_{\alpha_k} - x_{\alpha_{k-1}}}{\varepsilon} \right) - \frac{\partial V}{\partial x_{\alpha_k}} \zeta_{\alpha_k} \right]$$
$$\equiv A - \varepsilon \sum_{k=1}^{N+1} \frac{\partial V}{\partial x_{\alpha_k}} \zeta_{\alpha_k}. \quad \text{(C.6)}$$

Now, if we were working in the continuum, we would proceed with an integration by parts (see Section 3.3.3). Here we break up the sum and perform some changes in the indices using the fact that $\zeta_{\alpha_0} = \zeta_{\alpha_{N+1}} = 0$

$$A = \varepsilon \sum_{k=1}^{N+1} m \left(\frac{\zeta_{\alpha_k} - \zeta_{\alpha_{k-1}}}{\varepsilon} \right) \left(\frac{x_{\alpha_k} - x_{\alpha_{k-1}}}{\varepsilon} \right)$$
$$= m \sum_{k=1}^{N+1} \zeta_{\alpha_k} \left(\frac{x_{\alpha_k} - x_{\alpha_{k-1}}}{\varepsilon} \right) - m \sum_{k=1}^{N+1} \zeta_{\alpha_{k-1}} \left(\frac{x_{\alpha_k} - x_{\alpha_{k-1}}}{\varepsilon} \right)$$
$$= m \sum_{k=1}^{N+1} \zeta_{\alpha_k} \left(\frac{x_{\alpha_k} - x_{\alpha_{k-1}}}{\varepsilon} \right) - m \sum_{l=0}^{N} \zeta_{\alpha_l} \left(\frac{x_{\alpha_{l+1}} - x_{\alpha_l}}{\varepsilon} \right)$$
$$= m \sum_{k=1}^{N} \zeta_{\alpha_k} \left(\frac{x_{\alpha_k} - x_{\alpha_{k-1}}}{\varepsilon} \right) - m \sum_{l=1}^{N} \zeta_{\alpha_l} \left(\frac{x_{\alpha_{l+1}} - x_{\alpha_l}}{\varepsilon} \right)$$
$$= m \sum_{k=1}^{N} \zeta_{\alpha_k} \left(\frac{x_{\alpha_k} - x_{\alpha_{k-1}}}{\varepsilon} - \frac{x_{\alpha_{k+1}} - x_{\alpha_k}}{\varepsilon} \right)$$
$$= -m\varepsilon \sum_{k=1}^{N} \zeta_{\alpha_k} \left(\frac{x_{\alpha_{k-1}} - 2x_{\alpha_k} + x_{\alpha_{k+1}}}{\varepsilon^2} \right). \quad \text{(C.7)}$$

Hence

$$\delta S_N[x] = -\varepsilon \sum_{k=1}^{N} \zeta_{\alpha_k} \left[m \left(\frac{x_{\alpha_{k-1}} - 2x_{\alpha_k} + x_{\alpha_{k+1}}}{\varepsilon^2} \right) + \frac{\partial V}{\partial x_{\alpha_k}} \right]. \quad \text{(C.8)}$$

Clearly the expression in the squared brackets is vanishing when $x = x_{cl}$. In fact the discretized equation of motion reads

$$m\ddot{x} + \frac{\partial V}{\partial x} = 0 \xrightarrow[\text{discretized}]{} m\left(\frac{x_{\alpha_{k-1}} - 2x_{\alpha_k} + x_{\alpha_{k+1}}}{\varepsilon^2}\right) + \frac{\partial V}{\partial x_{\alpha_k}} = 0. \quad \text{(C.9)}$$

Therefore we have:

$$\delta S_N[x_{cl}] = 0. \quad \text{(C.10)}$$

Let us note that we have been able to change the summation from $\sum_{k=1}^{N+1}$ to $\sum_{k=1}^{N}$ thanks to the conditions for ζ at the endpoints. This is compatible with the presence of quantities having index α_{k+1}.

Now we consider the second variation of the action. This can be computed from Eqn.(C.8) with manipulations similar to those that we have just performed. By definition we have that

$$\delta^2 S_N[x] = \delta S_N[x + \zeta] - \delta S_N[x]. \quad \text{(C.11)}$$

Hence we find

$$\delta^2 S_N[x] = -\varepsilon \sum_{k=1}^{N} \zeta_{\alpha_k} \left[\frac{m}{\varepsilon^2}\left(\zeta_{\alpha_{k-1}} - 2\zeta_{\alpha_k} + \zeta_{\alpha_{k+1}}\right)\right.$$
$$\left. + \frac{\partial V}{\partial x_{\alpha_k}}\left(x_{\alpha_k} + \zeta_{\alpha_k}, t_{\alpha_k}\right) - \frac{\partial V}{\partial x_{\alpha_k}}\left(x_{\alpha_k}, t_{\alpha_k}\right)\right]. \quad \text{(C.12)}$$

Therefore, up to higher order terms, we get

$$\delta^2 S_N[x] = \sum_{k=1}^{N} \left[-\frac{m}{\varepsilon}\zeta_{\alpha_k}\left(\zeta_{\alpha_{k-1}} - 2\zeta_{\alpha_k} + \zeta_{\alpha_{k+1}}\right) - \varepsilon\frac{\partial^2 V}{\partial x_{\alpha_k}^2}\zeta_{\alpha_k}^2\right]$$
$$\equiv B - \varepsilon \sum_{k=1}^{N} \frac{\partial^2 V}{\partial x_{\alpha_k}^2}\zeta_{\alpha_k}^2. \quad \text{(C.13)}$$

We can rewrite the term B exploiting tricks similar to those used in Eqn.(C.7) and we obtain

$$
\begin{aligned}
B &= -\frac{m}{\varepsilon} \sum_{k=1}^{N} \zeta_{\alpha_k} \left(\zeta_{\alpha_{k-1}} - 2\zeta_{\alpha_k} + \zeta_{\alpha_{k+1}} \right) \\
&= -\frac{m}{\varepsilon} \sum_{k=1}^{N} \zeta_{\alpha_k} \left(\zeta_{\alpha_{k+1}} - \zeta_{\alpha_k} \right) + \frac{m}{\varepsilon} \sum_{k=1}^{N} \zeta_{\alpha_k} \left(\zeta_{\alpha_k} - \zeta_{\alpha_{k-1}} \right) \\
&= -\frac{m}{\varepsilon} \sum_{k=1}^{N} \zeta_{\alpha_k} \left(\zeta_{\alpha_{k+1}} - \zeta_{\alpha_k} \right) + \frac{m}{\varepsilon} \sum_{l=0}^{N-1} \zeta_{\alpha_{l+1}} \left(\zeta_{\alpha_{l+1}} - \zeta_{\alpha_l} \right) \\
&= -\frac{m}{\varepsilon} \sum_{k=1}^{N} \zeta_{\alpha_k} \left(\zeta_{\alpha_{k+1}} - \zeta_{\alpha_k} \right) + \frac{m}{\varepsilon} \sum_{l=0}^{N} \zeta_{\alpha_{l+1}} \left(\zeta_{\alpha_{l+1}} - \zeta_{\alpha_l} \right) \\
&= -\frac{m}{\varepsilon} \sum_{k=0}^{N} \zeta_{\alpha_k} \left(\zeta_{\alpha_{k+1}} - \zeta_{\alpha_k} \right) + \frac{m}{\varepsilon} \sum_{l=0}^{N} \zeta_{\alpha_{l+1}} \left(\zeta_{\alpha_{l+1}} - \zeta_{\alpha_l} \right) \\
&= \frac{m}{\varepsilon} \sum_{k=0}^{N} \left(\zeta_{\alpha_{k+1}} - \zeta_{\alpha_k} \right)^2 .
\end{aligned}
\tag{C.14}
$$

Changing the summation $\sum_{k=1}^{N}$ to $\sum_{k=0}^{N}$ in Eqn.(C.13) we finally have

$$
\delta^2 S_N[x] = \sum_{k=0}^{N} \left[\frac{m}{\varepsilon} \left(\zeta_{\alpha_{k+1}} - \zeta_{\alpha_k} \right)^2 - \varepsilon \frac{\partial^2 V}{\partial x_{\alpha_k}^2} \zeta_{\alpha_k}^2 \right].
\tag{C.15}
$$

Higher order variations are defined by the following formula

$$
\delta^n S_N[x] = \delta^{n-1} S_N[x + \zeta] - \delta^{n-1} S_N[x].
\tag{C.16}
$$

Since in Eqn.(C.15) the only dependence on x comes from the potential V, we obtain the following relation for $n > 2$:

$$
\delta^n S_N[x] = -\varepsilon \sum_{k=0}^{N} \frac{\partial^n V}{\partial x_{\alpha_k}^n} \zeta_{\alpha_k}^2 .
\tag{C.17}
$$

Appendix D

Brief review of Grassmann variables

Grassmann variables were introduced by Grassmann in order to describe *"ruled surfaces"* and were used much later in order to provide a classical analogue of spin [Casalbuoni (1976); Berezin and Marinov (1977)] and a path integral approach to systems with spin. They are usually indicated with the symbol ψ_i $(i = 1, \cdots, n)$ and are defined as a set of generators of the following algebra [Dewitt (1992)]:

$$\psi_i \psi_j + \psi_j \psi_i = 0$$

for $i, j = 1, \cdots, n$. This implies that $\psi_i^2 = 0$ and tells us that a generic function of ψ_i is just a finite sum:

$$f(\psi) = c + c_i \psi_i + c_{ij} \psi_i \psi_j + \cdots + c_{i_1 \cdots i_n} \psi_{i_1} \cdots \psi_{i_n}. \tag{D.1}$$

The inverse of such function exists only if the coefficient c is different from zero. For instance let us consider the case where we have only one Grassmann variable ψ, i.e.: $n = 1$. A generic function has the form

$$f(\psi) = a + b\psi$$

and its inverse is given by:

$$[f(\psi)]^{-1} = \frac{1}{a} - \frac{b}{a^2}\psi.$$

Note also that the coefficients appearing in Eqn.(D.1) are antisymmetric because of the anti-commuting nature of the variables ψ_i. Thus the expression (D.1) can naturally be interpreted as an n-form. This fact has been exploited in Section 5.3 to introduce a base for the differential forms on phase space by means of the Grassmann variables c^a and to interpret the charges associated to the CPI in a geometric way. Another meaning for

117

the Grassmann variables is that in field theory they are the eigenvalues of the fermionic destruction operators like the complex c-numbers z are the eigenvalue of the bosonic field destruction operator. More about this can be found in [Sakita (1985)].

In order to define differentiation with respect to a variable ψ_i we first bring Eqn.(D.1) to the form

$$f(\psi) = a + \psi_i b$$

where a and b generally depend on the other Grassmann variables. Differentiation is defined via the following relation

$$\frac{\partial}{\partial \psi_i}(a + \psi_i b) = b \,.$$

We immediately get that $(\partial/\partial\psi_i)^2 = 0$ which tells us that the operator $\partial/\partial\psi_i$ is nilpotent. In general we have

$$\frac{\partial}{\partial\psi_i}\psi_j + \psi_j\frac{\partial}{\partial\psi_i} = \delta_{ij}$$

and

$$\frac{\partial}{\partial\psi_i}\frac{\partial}{\partial\psi_j} + \frac{\partial}{\partial\psi_j}\frac{\partial}{\partial\psi_i} = 0 \,.$$

The integration is defined via the following two rules :

$$\int \mathrm{d}\psi\,\psi = 1$$

$$\int \mathrm{d}\psi\,c = 0$$

where c is a constant. We note that the above integration rules satisfy translational invariance:[1]

$$\int \mathrm{d}\psi\,\psi = \int \mathrm{d}\psi\,(\psi + c) \,.$$

For simplicity let us consider integration over a single Grassmann variable of the function $f(\psi) = a + b\psi$. If we perform a change of variables $\psi' = c\psi + d$ we have

$$\int \mathrm{d}\psi\,f(\psi) = \int \mathrm{d}\psi'\,J\,f\left(\frac{\psi' - d}{c}\right) \qquad (\mathrm{D.2})$$

---

[1]This is analogous to the shift invariance for the integral $\int_{-\infty}^{+\infty} \mathrm{d}x f(x+y) = \int_{-\infty}^{+\infty} \mathrm{d}x f(x)$.

where J is the Jacobian. The l.h.s. of Eqn.(D.2) follows directly from the integration rules and we have

$$\int \mathrm{d}\psi\, f(\psi) = b\,.$$

The r.h.s. of Eqn.(D.2) is given by

$$\int \mathrm{d}\psi'\, J\, f\left(\frac{\psi' - d}{c}\right) = \int \mathrm{d}\psi'\, J\left[a + \frac{b}{c}\,(\psi' - d)\right]$$

$$= J\frac{b}{c}\,.$$

This implies that $J = c$ which can be rewritten as

$$J^{-1} = \left(\frac{\partial \psi}{\partial \psi'}\right)\,.$$

Thus we note that the Grassmann Jacobian is just the inverse of the standard one. In the general case of n Grassmann variables we have

$$J^{-1} = \det\left(\frac{\partial \psi_i}{\partial \psi'_j}\right)\,.$$

So far we have been dealing with real Grassmann variables but our discussion can be straightforwardly extended to the complex case. In particular we denote with $\bar{\psi}_i$ the independent Grassmann variable defined as the conjugate element of ψ_i. The complex conjugation operation is defined (see [Dewitt (1992)]) as an operation such that the complex conjugation of $\bar{\psi}_i$ is ψ_i. The extension of the notion of integration is given by the following measure

$$\mathrm{d}\psi\,\mathrm{d}\bar{\psi} \equiv \prod_{i=1}^{n} \mathrm{d}\psi_i\,\mathrm{d}\bar{\psi}_i\,.$$

Gaussian integrals over complex Grassmann variables appear frequently in path integrals. Thus we end this appendix by computing such integrals in an important case. Let us consider:

$$\int \mathrm{d}\psi\,\mathrm{d}\bar{\psi}\, e^{\sum_{ij} \bar{\psi}_i A_{ij} \psi_j} = \sum_{m=1}^{n} \int \mathrm{d}\psi\,\mathrm{d}\bar{\psi}\,\frac{1}{m!}\left(\sum_{ij} \bar{\psi}_i A_{ij} \psi_j\right)^m\,.$$

The only term contributing to the integral is the one where all the Grassmann variables appear. Hence we have

$$\int \mathrm{d}\psi\,\mathrm{d}\bar{\psi}\, e^{\sum_{ij} \bar{\psi}_i A_{ij} \psi_j} = \frac{1}{n!}\int \prod_{k=1}^{n} \mathrm{d}\psi_k\,\mathrm{d}\bar{\psi}_k\left(\sum_{ij} \bar{\psi}_i A_{ij} \psi_j\right)^n$$

Path Integrals for Pedestrians

After some manipulations we obtain

$$\left(\sum_{ij} \bar{\psi}_i A_{ij} \psi_j \right)^n = \bar{\psi}_1 \psi_1 \cdots \bar{\psi}_n \psi_n \varepsilon_{j_1 \cdots j_n} \varepsilon_{i_1 \cdots i_n} A_{i_1 j_1} \cdots A_{i_n j_n} ,$$

and recalling that

$$\det (A) = \frac{1}{n!} \varepsilon_{j_1 \cdots j_n} \varepsilon_{i_1 \cdots i_n} A_{i_1 j_1} \cdots A_{i_n j_n}$$

we find

$$\int \mathrm{d}\psi \, \mathrm{d}\bar{\psi} e^{\sum_{ij} \bar{\psi}_i A_{ij} \psi_j} = \det (A) .$$

Note that the result is the inverse of the one we would get if the ψ_i were complex c-numbers.

Appendix E

Dimensional analysis of θ and $\bar{\theta}$

In this appendix we will analyze the complex nature and the physical dimensions of the Grassmann variables c, $\bar{c}$, θ, $\bar{\theta}$ entering the definition (5.59) of the superfield Φ^a:

$$\Phi^a = \varphi^a + \theta c^a + \bar{\theta}\omega^{ab}\bar{c}_b + i\bar{\theta}\theta\omega^{ab}\lambda_b.$$

Let us consider the superfield Φ^a as an operator $\widehat{\Phi}^a$ and impose the scalar product under which the Grassmann operators $\widehat{c}$ and $\widehat{\bar{c}}$ are Hermitian, see [Deotto et al. (2003a,b)] for further details. Since the first component $\widehat{\varphi}$ of the superfield $\widehat{\Phi}$ is Hermitian we require the entire superfield operator to be Hermitian. From

$$\widehat{\Phi}^{a\dagger} = \widehat{\Phi}^a$$

$$\Updownarrow$$

$$\widehat{\varphi}^a + \widehat{c}^a\theta^* + \omega^{ab}\widehat{\bar{c}}_b\bar{\theta}^* - i\omega^{ab}\widehat{\lambda}_b\theta^*\bar{\theta}^* = \widehat{\varphi}^a + \theta\widehat{c}^a + \bar{\theta}\omega^{ab}\widehat{\bar{c}}_b + i\bar{\theta}\theta\omega^{ab}\widehat{\lambda}_b$$

we easily get that the two Grassmann partners of time θ and $\bar{\theta}$ must be *imaginary*:

$$\theta^* = -\theta, \qquad \bar{\theta}^* = -\bar{\theta}. \qquad (E.1)$$

Of course, other choices of the scalar product imply other conventions about the character of the variables θ and $\bar{\theta}$. For example, if we impose the symplectic scalar product [Deotto et al. (2003a,b)], under which the Hermiticity conditions among the Grassmann operators are given by:

$$(\widehat{c}^a)^\dagger = i\omega^{ab}\widehat{c}_b, \qquad (\widehat{\bar{c}}_a)^\dagger = i\omega_{ab}\widehat{c}^b,$$

and we require the superfield to be Hermitian:

$$\widehat{\Phi}^{a\dagger} = \widehat{\Phi}^a$$

$$\Updownarrow$$

$$\widehat{\varphi}^a + i\omega^{ab}\widehat{\bar{c}}_b\theta^* + i\widehat{c}^a\bar{\theta}^* - i\theta^*\bar{\theta}^*\omega^{ab}\widehat{\lambda}_b = \widehat{\varphi}^a + \theta\widehat{c}^a + \bar{\theta}\omega^{ab}\widehat{\bar{c}}_b + i\bar{\theta}\theta\omega^{ab}\widehat{\lambda}_b$$

we get the following relations among the Grassmann partners of time:

$$\bar{\theta}^* = i\theta, \qquad \theta^* = i\bar{\theta}.$$

Finally, with the scalar product under which $\hat{c}^{a\dagger} = \hat{\bar{c}}_a$ [Deotto *et al.* (2003a,b)], it is impossible to have a Hermitian superfield operator. Anyhow, if we stick to the scalar product under which the operators $\hat{c}$ and $\hat{\bar{c}}$ are Hermitian, we have to take the Grassmann partners of time θ and $\bar{\theta}$ to be imaginary, i.e. Eqn.(E.1). With this choice it is easy to prove that the presence of the factor "i" in Eqn.(5.86) is crucial in order to have a real integration measure:

$$(i\,d\theta\,d\bar{\theta})^* = -i\,d\bar{\theta}^*\,d\theta^* = -i(-\,d\bar{\theta})(-\,d\theta) = -i\,d\bar{\theta}\,d\theta = i\,d\theta\,d\bar{\theta}.$$

Not only that, but with the scalar product under which the Grassmann operators $\hat{c}$ and $\hat{\bar{c}}$ are Hermitian, the BRS and anti-BRS charge of Eqn.(5.40) are anti-Hermitian, i.e. $\hat{Q}^\dagger = -\hat{Q}$ and $\hat{\bar{Q}}^\dagger = -\hat{\bar{Q}}$. Consequently, the operators $\exp(\theta\hat{Q})$ and $\exp(\hat{\bar{Q}}\bar{\theta})$, entering the connection between the Schrödinger and the Heisenberg pictures given by Eqn.(5.63), are unitary. In fact the anti-Hermiticity of $\hat{Q}$, $\hat{\bar{Q}}$ and Eqn.(E.1) imply:

$$[\exp(\theta\hat{Q})]^\dagger = \exp(\hat{Q}^\dagger\theta^*) = \exp(\hat{Q}\theta) = \exp(-\theta\hat{Q}),$$

which means that $\exp(\theta\hat{Q})$ is a unitary operator. A similar proof holds for $\exp(\hat{\bar{Q}}\bar{\theta})$.

Another interesting point to discuss is the physical dimension of the Grassmann variables θ and $\bar{\theta}$. Let us start by considering that, in the definition (5.71) of the superfields, q and $\bar{\theta}\theta\lambda_p$ appear in the same multiplet. So, from the point of view of the physical dimensions, we will have:

$$[q] = [\bar{\theta}\theta][\lambda_p] \longrightarrow [\bar{\theta}\theta] = [q][\lambda_p]^{-1}.$$

Now $[p, \lambda_p] = i$, so the dimensions of λ_p are just the inverse of the dimensions of p and as a consequence we can derive that the product $\bar{\theta}\theta$ has the dimensions of an action:

$$[\bar{\theta}\theta] = [q][p] = [\hbar]. \tag{E.2}$$

The dimensions of the product of the infinitesimals $d\theta\,d\bar{\theta}$ instead can be derived from the following equation:

$$\widetilde{\mathcal{H}} = i\int d\theta\,d\bar{\theta}\,H.$$

As an operator $\widetilde{\mathcal{H}}$ is a derivative with respect to t, so it has the dimensions of the inverse of a time: $[\widetilde{\mathcal{H}}] = T^{-1}$ while H is an energy, so its dimensions are: $[H] = ML^2T^{-2}$. This immediately implies that the dimensions of $d\theta \, d\bar\theta$ are given by:

$$[d\theta \, d\bar\theta] = [\tilde{\mathcal{H}}]/[H] = M^{-1}L^{-2}T = L^{-1}(LMT^{-1})^{-1} = [q]^{-1}[p]^{-1}.$$

From the previous equation we derive that the product $d\theta \, d\bar\theta$ has the dimensions of the inverse of an action:

$$[d\theta \, d\bar\theta] = [\hbar]^{-1}. \tag{E.3}$$

The previous equation is consistent with the standard definition of the Grassmann integration, which is the following one:

$$\int d\theta \, 1 = 0, \qquad \int d\bar\theta \, 1 = 0, \qquad \int d\theta \, \theta = 1, \qquad \int d\bar\theta \, \bar\theta = 1.$$

Such definition implies that $[d\theta] = [\theta]^{-1}$ and $[d\bar\theta] = [\bar\theta]^{-1}$, i.e. the dimensions of the infinitesimal of a Grassmann variable are just the inverse of the dimensions of the Grassmann variable itself. This is perfectly consistent with Eqns (E.2) and (E.3).

Even if the dimensions of the product $\theta\bar\theta$ are uniquely determined, there is still an arbitrariness in the dimensions of the single variables θ and $\bar\theta$. In fact, from the expression of the superfields:

$$Q = q + \theta c^q + \bar\theta \bar c_p + i\bar\theta\theta\lambda_p, \qquad P = p + \theta c^p - \bar\theta \bar c_q - i\bar\theta\theta\lambda_q,$$

and the fact that all the components of a superfield must have the same physical dimensions, we can derive that q must have the same physical dimensions as θc^q and $\bar\theta \bar c_p$, while p must have the same physical dimensions as θc^p and $\bar\theta \bar c_q$:

$$[q] = [\theta c^q] = [\bar\theta \bar c_p] \tag{E.4}$$

$$[p] = [\theta c^p] = [\bar\theta \bar c_q]. \tag{E.5}$$

Nevertheless we cannot determine in a unique way the dimensions of the single fields c^q, c^p, $\bar c_q$ and $\bar c_p$. The only thing that we can derive, dividing Eqn.(E.4) by Eqn.(E.5), is that the following equations hold:

$$\frac{[c^q]}{[c^p]} = \frac{[q]}{[p]}, \qquad \frac{[\bar c_q]}{[\bar c_p]} = \frac{[p]}{[q]}. \tag{E.6}$$

This means that only the ratio of the dimensions of the variables c and the ratio of the dimensions of the variables $\bar c$ can be determined. It is quite easy to show that Eqns.(E.2) and (E.6) guarantee the dimensional consistency of all the formulae that can be derived from the CPI. In any case, we must stress the fact that there is nothing in the theory that can fix the dimensions of the single Grassmann variables c and $\bar{c}$. Consequently there is nothing in the theory that can fix the dimensions of the single Grassmann partners of time θ and $\bar{\theta}$.

Anyhow since the Grassmann variables c can be interpreted as the Jacobi fields or the first variations of the theory $c^a \approx \delta\varphi^a$ we could take the physical dimensions of c^q to be identical to the physical dimensions of q and the physical dimensions of c^p to be identical to the physical dimensions of p. With these conventions we would have that θ is dimensionless while $\bar{\theta}$, in order to satisfy Eqn. (E.2), must have the dimensions of an action.

Appendix F

Schrödinger and Heisenberg picture in θ and $\bar{\theta}$

In standard quantum mechanics an explicit dependence of the operators on time t can be introduced by passing from the Schrödinger picture to the Heisenberg one via the following formula:

$$A_H(t) = e^{iHt/\hbar} A_S e^{-iHt/\hbar}.$$

Let us notice that in the argument of the exponential the time parameter t is coupled with the operator that generates the time evolution H.

As we said in Section 5.4.1, we can have an Heisenberg picture also with respect to θ and $\bar{\theta}$. The analog of the translation operator in t are now the translation operators in θ and $\bar{\theta}$. So, for example, we can go to the Heisenberg picture of the field φ^a. In Chapter 5 we said that the result is

$$\exp[\theta Q + \bar{Q}\bar{\theta}] \, \varphi^a \, \exp[-\theta Q - \bar{Q}\bar{\theta}] = \Phi^a(\theta, \bar{\theta}) \tag{F.1}$$

and we are now going to prove it. Let us notice that Q and $\bar{Q}$ are two commuting operators. Therefore the Baker-Campbell-Hausdorff formula gives for the first term on the r.h.s. of Eqn.(F.1):

$$\exp[\theta Q + \bar{Q}\bar{\theta}] = \exp[\theta Q] \exp[\bar{Q}\bar{\theta}] = (1 + \theta Q)(1 + \bar{Q}\bar{\theta})$$

and the whole l.h.s. of Eqn.(F.1) becomes:

$$\begin{aligned}
e^{\theta Q + \bar{Q}\bar{\theta}} \varphi^a e^{-\theta Q - \bar{Q}\bar{\theta}} &= (1 + \theta Q)(1 + \bar{Q}\bar{\theta})\varphi^a(1 - \theta Q)(1 - \bar{Q}\bar{\theta}) \\
&= \varphi^a + \theta[Q, \varphi^a] - \bar{\theta}[\bar{Q}, \varphi^a] + \bar{\theta}\theta[-\varphi^a Q\bar{Q} - \bar{Q}\varphi^a Q + Q\varphi^a\bar{Q} - Q\bar{Q}\varphi^a] \\
&= \varphi^a + \theta c^a - \bar{\theta}\omega^{ba}\bar{c}_b + \bar{\theta}\theta[[Q, \varphi^a], \bar{Q}] = \varphi^a + \theta c^a - \bar{\theta}\omega^{ba}\bar{c}_b + \bar{\theta}\theta[c^a, \bar{Q}] \\
&= \varphi^a + \theta c^a + \bar{\theta}\omega^{ab}\bar{c}_b + i\bar{\theta}\theta\omega^{ab}\lambda_b \\
&= \Phi^a. \tag{F.2}
\end{aligned}$$

This proves Eqn.(F.1). So we can look at the superfield operator Φ^a as the Heisenberg picture of the operator φ^a, with respect to the Grassmann partners of time θ and $\bar{\theta}$.

The procedure analyzed above can be applied also to a generic function $F(\varphi)$. Let us define the operator:

$$S \equiv \exp\left[\theta Q + \bar{Q}\bar{\theta}\right]$$

and let us see which type of transformation it induces on a generic function $F(\varphi)$. We will have:

$$
\begin{aligned}
SFS^{-1} &= \exp\left[\theta Q + \bar{Q}\bar{\theta}\right] F \exp\left[-\theta Q - \bar{Q}\bar{\theta}\right] \\
&= F + \theta[Q,F] - \bar{\theta}[\bar{Q},F] + \bar{\theta}\theta\left[[Q,F]\bar{Q} + \bar{Q}[Q,F]\right] \\
&= F + \theta c^a \partial_a F - \bar{\theta}\bar{c}_a \omega^{ab} \partial_b F + \bar{\theta}\theta\left[\bar{Q}, c^a \partial_a F\right] \\
&= F + \theta c^a \partial_a F - \bar{\theta}\bar{c}_a \omega^{ab} \partial_b F + \bar{\theta}\theta(-i\lambda_a \omega^{ab}\partial_b F + \bar{c}_a \omega^{ab}\partial_b \partial_d F c^d).
\end{aligned}
$$

In the particular case in which the function F is just the Hamiltonian H we get that:

$$SHS^{-1} = H + \theta N + \bar{N}\bar{\theta} - i\bar{\theta}\theta\widetilde{\mathcal{H}}, \qquad \text{(F.3)}$$

i.e., the multiplet SHS^{-1} includes both the Hamiltonian H that appears in the weight of the quantum path integral and the Hamiltonian $\widetilde{\mathcal{H}}$ that appears in the weight of the classical path integral, besides the two conserved charges N and $\bar{N}$. Eqn.(F.3) can also be written in a compact form as:

$$SH(\varphi)S^{-1} = H(\Phi), \qquad \text{(F.4)}$$

which means that the Heisenberg picture of the operator $H(\varphi)$ is just given by the same function H but with the fields φ replaced by the superfields Φ.

To prove that the r.h.s. of Eqn.(F.3) is just given by $H(\Phi)$ let us evaluate explicitly the expansion of $H(\Phi^a) = H(\varphi^a + X^a)$ where $X^a \equiv \theta c^a + \bar{\theta}\omega^{ab}\bar{c}_b + i\bar{\theta}\theta\omega^{ab}\lambda_b$. Using the properties of the Grassmann variables, the expansion of the Hamiltonian $H(\Phi)$ becomes:

$$
\begin{aligned}
H(\Phi) &= H(\varphi) + X^a \partial_a H + \frac{1}{2}X^a X^b \partial_a \partial_b H \\
&= H(\varphi) + \theta c^a \partial_a H + \bar{\theta}\omega^{ab}\bar{c}_b \partial_a H + i\bar{\theta}\theta\omega^{ab}\lambda_b \partial_a H \\
&\quad + \frac{1}{2}(\theta c^a + \bar{\theta}\omega^{ae}\bar{c}_e)(\theta c^b + \bar{\theta}\omega^{bf}\bar{c}_f)\partial_a \partial_b H \\
&= H(\varphi) + \theta N - \bar{\theta}\bar{N} - i\bar{\theta}\theta(\lambda_a \omega^{ab}\partial_b H + i\bar{c}_a \omega^{ab}\partial_b \partial_d H c^d),
\end{aligned}
$$

which is just the r.h.s. of Eqn.(F.3). From Eqns.(F.3) and (F.4) we can derive that the relationship between H and $\widetilde{\mathcal{H}}$ is given by:

$$i\int d\theta\, d\bar{\theta}\, H(\Phi) = \widetilde{\mathcal{H}}. \qquad \text{(F.5)}$$

A similar formula holds also for the Lagrangian. In particular, let us consider the Lagrangian

$$L = p\dot{q} - H(q,p). \tag{F.6}$$

If we replace the fields with the superfields in the kinetic term $p\dot{q}$ and we integrate over θ, $\bar{\theta}$, what we get is

$$i \int \mathrm{d}\theta\,\mathrm{d}\bar{\theta}\ \Phi^p \dot{\Phi}^q = i \int \mathrm{d}\theta\,\mathrm{d}\bar{\theta}(p + \theta c^p - \bar{\theta}\bar{c}_q - i\bar{\theta}\theta\lambda_q)(\dot{q} + \theta\dot{c}^q + \bar{\theta}\dot{\bar{c}}_p + i\bar{\theta}\theta\dot{\lambda}_p)$$

$$= \lambda_q \dot{q} + i\bar{c}_q \dot{c}^q + ic^p \dot{\bar{c}}_p - p\dot{\lambda}_p$$

$$= \lambda_a \dot{\varphi}^a + i\bar{c}_a \dot{c}^a - \frac{d}{dt}(\lambda_p p + i\bar{c}_p c^p). \tag{F.7}$$

Collecting together Eqns.(F.5), (F.6) and (F.7) we get exactly Eqn. (5.70):

$$i \int \mathrm{d}\theta\,\mathrm{d}\bar{\theta}L(\Phi) = \widetilde{\mathcal{L}} - \frac{d}{dt}(\lambda_p p + i\bar{c}_p c^p).$$

Appendix G

Classical path integral in the momentum representation

In this appendix we will give some details about the classical and the quantum path integrals in the momentum representation. First of all, we want to give some details on the derivation of Eqn.(5.90). Let us start analyzing the weight of the path integral. From Eqn.(5.74) we know that the weight appearing in the path integral of the kernel $\langle q, p, c^q, c^p, t | q_0, p_0, c^{q_0}, c^{p_0}, t_0 \rangle$ contains a Lagrangian $L[\Phi]$ plus some surface terms (5.75) of the form:

$$(\text{s.t.}) = i\lambda_p p - i\lambda_{p_0} p_0 - \bar{c}_p c^p + \bar{c}_{p_0} c^{p_0}. \tag{G.1}$$

When we insert the expression of the kernel (5.74) into Eqn.(5.89) we see that we must consider, besides the terms of Eqn. (G.1), also the phase factors coming from the Fourier transforms (5.89). These terms can be collected together and written in terms of the superfields as follows:

$$\exp\left[i\lambda_p p - i\lambda_{p_0} p_0 - \bar{c}_p c^p + \bar{c}_{p_0} c^{p_0} - i\lambda_q q + \bar{c}_q c^q + i\lambda_{q_0} q_0 - \bar{c}_{q_0} c^{q_0} \right]$$

$$= \exp\left[-i \int i\,d\theta\,d\bar{\theta}(q + \theta c^q + \bar{\theta}\bar{c}_p + i\bar{\theta}\theta\lambda_p)(p + \theta c^p - \bar{\theta}\bar{c}_q - i\bar{\theta}\theta\lambda_q) \right]\Bigg|_{t_0}^{t}$$

$$= \exp\left[-i \int_{t_0}^{t} i\,d\tau\,d\theta\,d\bar{\theta}\,\frac{d(QP)}{d\tau} \right]. \tag{G.2}$$

This proves that the surface terms appearing in the weight of Eqn.(5.90) are the right ones.

If we want now to prove that, in going from Eqn.(5.77) to Eqn.(5.90), the functional measure changes from $\mathcal{D}''Q\mathcal{D}P$ to $\mathcal{D}Q\mathcal{D}''P$ then we must perform the same steps that are usually done in quantum mechanics to prove that the functional measure changes from $\mathcal{D}''q\mathcal{D}p$ to $\mathcal{D}q\mathcal{D}''p$ in passing from the path integral in the coordinate representation (5.84) to the one in the momenta (5.93). In particular, in order to understand the meaning of the functional measures $\mathcal{D}''q\mathcal{D}p$ and $\mathcal{D}q\mathcal{D}''p$, we must come back to the discretized versions of the quantum path integrals as we did in Chapter 2. Let us start with the one in the coordinate representation $\langle q, t | q_0, t_0 \rangle = \langle q | \left[\exp(-\frac{i}{\hbar}(t - t_0)H) \right] | q_0 \rangle$. If we slice the time interval $t - t_0$ in N intervals of length ϵ and we use the formula $\exp \left[-\frac{i}{\hbar}(t - t_0)H \right] = \exp \left(-\frac{i}{\hbar}\epsilon H \right)^N$ then we can insert a completeness relation in q before every exponential factor and a completeness in p after every exponential factor to get the following equation [Kleinert (1990)]:

$$\langle q, t | q_0, t_0 \rangle = \int \frac{dp_N}{2\pi\hbar} dq_{N-1} \frac{dp_{N-1}}{2\pi\hbar} \cdots dq_1 \frac{dp_1}{2\pi\hbar} \langle q | \exp\left(-i\epsilon H/\hbar\right) | p_N \rangle \langle p_N | q_{N-1} \rangle$$
$$\times \langle q_{N-1} | \exp\left(-i\epsilon H/\hbar\right) | p_{N-1} \rangle \cdots \langle q_1 | \exp\left(-i\epsilon H/\hbar\right) | p_1 \rangle \langle p_1 | q_0 \rangle$$
$$= \lim_{N\to\infty} \prod_{n=1}^{N-1} \left[\int_{-\infty}^{\infty} dq_n \right] \prod_{n=1}^{N} \left[\int_{-\infty}^{\infty} \frac{dp_n}{2\pi\hbar} \right] \exp \left[\frac{i}{\hbar} \mathscr{A}^N \right]$$

with $\mathscr{A}^N = \sum_{n=1}^{N} \left[p_n(q_n - q_{n-1}) - \epsilon H(p_n, q_n) \right]$. In the previous path integral the initial coordinate q_0 and the final one $q \equiv q_N$ are not integrated over, but they are fixed by the boundary conditions. All the momenta, from the initial p_1 to the final p_N, are instead integrated over: this is the meaning of the functional measure $\mathcal{D}''q\mathcal{D}p$. Now, to go from the path integral in the coordinate representation to the one in the momenta, we must perform a Fourier transform over the initial and final coordinates:

$$\langle p, t | p_0, t_0 \rangle = \int dq \, dq_0 \, \exp \left[-\frac{i}{\hbar} pq \right] \langle q, t | q_0, t_0 \rangle \exp \left[\frac{i}{\hbar} p_0 q_0 \right].$$

Also in this case we can slice the time interval $t - t_0$ into N intervals of length ϵ and insert the completeness relations. This time we will insert the completeness relations in q after each exponential factor $\exp\left(-i\epsilon H/\hbar\right)$ and those in p before it. What we get is:

$$\int dq_N \frac{dp_{N-1}}{2\pi\hbar} dq_{N-1} \cdots \frac{dp_1}{2\pi\hbar} dq_1 \langle p | \exp\left(-i\epsilon H/\hbar\right) | q_N \rangle \langle q_N | p_{N-1} \rangle$$
$$\times \langle p_{N-1} | \left(-i\epsilon H/\hbar\right) | q_{N-1} \rangle \cdots \langle p_1 | \exp\left(-i\epsilon H/\hbar\right) | q_1 \rangle \langle q_1 | p_0 \rangle$$
$$= \lim_{N\to\infty} \prod_{n=1}^{N} \left[\int_{-\infty}^{\infty} dq_n \right] \prod_{n=1}^{N-1} \left[\int_{-\infty}^{\infty} \frac{dp_n}{2\pi\hbar} \right] \exp \left[\frac{i}{\hbar} \mathscr{B}^N \right]. \tag{G.3}$$

with $\mathscr{B}^N = \sum_{n=1}^{N} \left[-q_n(p_n - p_{n-1}) - \epsilon H(p_n, q_n) \right].$

In the path integral (G.3) the initial and the final momenta p_0 and $p \equiv p_N$ are fixed, while all the coordinates from q_1 to q_N are integrated over. In the continuum limit Eqn.(G.3) reproduces the path integral (5.93), i.e.:

$$\langle p, t | p_0, t_0 \rangle = \int \mathcal{D}q \mathcal{D}''p \, \exp\left[\frac{i}{\hbar} \int_{t_0}^{t} \mathrm{d}\tau \left[-q\dot{p} - H(q, p, t)\right]\right].$$

The final result of these manipulations in the continuum limit was to change the kinetic terms of the quantum path integral from $p\dot{q}$ to $-q\dot{p}$. These kinetic terms can be connected via the surface term $\frac{d}{dt}(qp)$, according to the formula: $-q\dot{p} = p\dot{q} - \frac{d}{dt}(qp)$. As we have seen in Eqn.(G.2) this connection becomes $-Q\dot{P} = P\dot{Q} - \frac{d}{dt}(QP)$ at the classical path integral level.

Appendix H

Classical path integral via the Trotter formula

In this appendix we want to prove that, like in standard quantum theory, it is possible to derive the classical path integral (5.85) using, from the beginning, the completeness relations and the Trotter formula. Next we will write the graded commutators given bu Eqns.(5.31) through (5.33), which characterize the operatorial theory associated to the CPI, in terms of the superfields.

First of all let us remember that we can define the fundamental kets $|Q\rangle$ and $|P\rangle$ via the relations:

$$\begin{cases} \widehat{Q}(\theta,\bar{\theta})|Q\rangle = Q(\theta,\bar{\theta})|Q\rangle \\ \widehat{P}(\theta,\bar{\theta})|P\rangle = P(\theta,\bar{\theta})|P\rangle. \end{cases} \tag{H.1}$$

From Eqn.(H.1) the fundamental kets $|Q\rangle$ and $|P\rangle$ are uniquely determined. They are the eigenstates of the operators $\widehat{Q}(\theta,\bar{\theta})$ and $\widehat{P}(\theta,\bar{\theta})$ with eigenvalues $Q(\theta,\bar{\theta})$ and $P(\theta,\bar{\theta})$. With this definition we can identify

$$|Q\rangle \equiv |q, \lambda_p, c^q, \bar{c}_p\rangle, \qquad |P\rangle \equiv |\lambda_q, p, \bar{c}_q, c^p\rangle. \tag{H.2}$$

If we assume that integrating over a superfield is equivalent to integrating over all its components then the completeness relations can be written in a compact form as:

$$\int dQ|Q\rangle\langle Q| = \mathbb{I}, \qquad \int dP|P\rangle\langle P| = \mathbb{I},$$

where $dQ \equiv dq\, d\lambda_p\, dc^q\, d\bar{c}_p$ and $dP \equiv d\lambda_q\, dp\, d\bar{c}_q\, dc^p$. Now we can derive the CPI in terms of the superfields via standard techniques, such as the Trotter formula, the completeness relations, and so on. The starting point is given, as usual, by the operator $\exp\left[-i(t-t_0)\widehat{\widetilde{\mathcal{H}}}\right]$ that can be rewritten as $\exp\left[-i(t-t_0)\int i\,d\theta\,d\bar{\theta}\,H(\widehat{\Phi})\right]$. To get the associated path integral we

133

must sandwich the previous expression between $\langle Q, t|$ and $|Q_0, t_0\rangle$. If we divide the time interval $t - t_0$ in N intervals of length ϵ and we use the formula

$$\exp\left[-i(t - t_0)\widehat{\widetilde{\mathcal{H}}}\right] = \left(\exp\left[-i\epsilon\widehat{\widetilde{\mathcal{H}}}\right]\right)^N$$

then we get N exponential factors $\exp\left[-i\epsilon\widehat{\widetilde{\mathcal{H}}}\right]$. Inserting a completeness relation in Q before every exponential factor and a completeness relation in P after it we easily obtain:

$$\langle Q, t|Q_0, t_0\rangle = \lim_{N\to\infty} \int \prod_{j=1}^{N-1} \mathrm{d}Q_j \prod_{j=1}^{N} \mathrm{d}P_j \prod_{j=1}^{N} \mathscr{A}_{j,j-1},$$

where $\mathscr{A}_{j,j-1}$ is given by

$$\mathscr{A}_{j,j-1} = \langle P_j|Q_{j-1}\rangle\langle Q_j|\exp\left[-i\epsilon\widehat{\widetilde{\mathcal{H}}}\right]|P_j\rangle =$$

$$= \langle P_j|Q_{j-1}\rangle\langle Q_j|\exp\left[-i\epsilon\int i\,\mathrm{d}\theta\,\mathrm{d}\bar{\theta}H(\widehat{Q},\widehat{P})\right]|P_j\rangle.$$

Using the definition (H.2) of the states $|Q\rangle$ and $|P\rangle$ it is easy to prove that the scalar product between them can also be written in terms of the superfields: $\langle P|Q\rangle = \exp i\int i\,\mathrm{d}\theta\,\mathrm{d}\bar{\theta}(-PQ)$. Consequently we can rewrite $\mathscr{A}_{j,j-1}$ as:

$$\mathscr{A}_{j,j-1} = \exp\left[i\int i\,\mathrm{d}\theta\,\mathrm{d}\bar{\theta}\left(-P_jQ_{j-1} + P_jQ_j - \epsilon H[\Phi_j]\right)\right] =$$

$$= \exp\left[i\epsilon\int i\,\mathrm{d}\theta\,\mathrm{d}\bar{\theta}\left(P_j\frac{Q_j - Q_{j-1}}{\epsilon} - H[\Phi_j]\right)\right].$$

In the limit $N \to \infty$ and $\epsilon \to 0$ we get just the classical path integral (5.85) written in terms of the superfields:

$$\langle Q, t|Q_0, t_0\rangle = \int \mathcal{D}''Q(\theta, \bar{\theta})\mathcal{D}P(\theta, \bar{\theta}) \exp\left[i\int_{t_0}^{t} i\,\mathrm{d}\tau\,\mathrm{d}\theta\,\mathrm{d}\bar{\theta}\left(P(\theta, \bar{\theta})\dot{Q}(\theta, \bar{\theta})\right.\right.$$

$$\left.\left. - H(\Phi(\theta, \bar{\theta}))\right)\right]. \tag{H.3}$$

As we wrote in Chapter 5, from the previous expression it is possible to derive the graded commutators of Eqns. (5.31) through (5.33) and to write them in terms of the superfields. To do this, we can use an analogy between the CPI and the usual quantum *field* theories. For example the quantum path integral in phasse space for a scalar field $\phi(\mathbf{x})$ is given by:

$$\int \mathcal{D}''\phi(\mathbf{x})\mathcal{D}\pi(\mathbf{x}) \ \exp\frac{i}{\hbar}\int dt\,d\mathbf{x}\,\left[\pi(\mathbf{x})\dot{\phi}(\mathbf{x}) - \mathbf{H}(\mathbf{x})\right]$$

and it is equivalent to an operatorial theory in which the only non-zero equal time commutators are:

$$[\phi(\mathbf{x},t),\pi(\mathbf{y},t)] = i\hbar\delta(\mathbf{x}-\mathbf{y}). \tag{H.4}$$

Which is the operatorial theory lying behind the classical path integral (H.3)? If we look at $Q(\theta,\bar{\theta})$ and $P(\theta,\bar{\theta})$ as fields in which the space variables $\mathbf{x}$ are replaced by the Grassmann variables θ and $\bar{\theta}$, we expect that the equal time commutator satisfied by Q and P, analogous to Eqn.(H.4), is the following one:

$$[Q(t,\theta,\bar{\theta}),P(t,\theta',\bar{\theta}')] = \delta(\bar{\theta}-\bar{\theta}')\delta(\theta-\theta'). \tag{H.5}$$

Let us see whether, using the definition of the superfields

$$Q(t,\theta,\bar{\theta}) = q(t) + \theta c^q(t) + \bar{\theta}\bar{c}_p(t) + i\bar{\theta}\theta\lambda_p(t)$$

$$P(t,\theta',\bar{\theta}') = p(t) + \theta' c^p(t) - \bar{\theta}'\bar{c}_q(t) - i\bar{\theta}'\theta'\lambda_q(t),$$

Eqn. (H.5) leads to the commutators of Eqns.(5.31) through (5.33) of the CPI. Because of the properties of the Grassmannian Dirac deltas, Eqn.(H.5) can be rewritten as

$$[Q(\theta,\bar{\theta}),P(\theta',\bar{\theta}')] = \bar{\theta}\theta - \bar{\theta}\theta' - \bar{\theta}'\theta + \bar{\theta}'\theta'. \tag{H.6}$$

Expanding the l.h.s. of Eqn. (H.6) in terms of θ, $\bar{\theta}$, θ' and $\bar{\theta}'$ we get:

$$[Q(\theta,\bar{\theta}),P(\theta',\bar{\theta}')] = -i\bar{\theta}'\theta'\,[q,\lambda_q]+\theta\bar{\theta}'\,[c^q,\bar{c}_q]-\bar{\theta}\theta'\,[\bar{c}_p,c^p]+i\bar{\theta}\theta\,[\lambda_p,p]+\cdots. \tag{H.7}$$

Comparing Eqns.(H.6) and (H.7) we obtain that the non-zero graded commutators are:

$$[q,\lambda_q] = i, \qquad [c^q,\bar{c}_q] = 1, \qquad [\bar{c}_p,c^p] = 1, \qquad [\lambda_p,p] = -i,$$

which are precisely the graded commutators of Eqns.(5.31) through (5.33) of the CPI. Of course, since on the l.h.s. of Eqn.(H.7) we have a commutator and not an anticommutator, we get that:

$$[P(t,\theta,\bar{\theta}),Q(t,\theta',\bar{\theta}')] = -\delta(\bar{\theta}-\bar{\theta}')\delta(\theta-\theta'),$$

while $[Q,Q] = [P,P] = 0$. Summarizing, we can say that the graded commutators of the CPI can be written in a compact form via the superfields, as follows:

$$[\Phi^a(t,\theta,\bar{\theta}),\Phi^b(t,\theta',\bar{\theta}')] = \omega^{ab}\delta(\bar{\theta}-\bar{\theta}')\delta(\theta-\theta').$$

Appendix I

Ordering problems in the classical path integral

In this appendix we want to analyze the issue of ordering problems in the CPI. In quantum mechanics the ordering problems arise because $\hat{q}$ and $\hat{p}$ do not commute, so there is more than one Hermitian operator $H(\hat{q},\hat{p})$ associated with the same classical Hamiltonian $H(q,p)$. For this reason we must specify the order in which we consider the operators $\hat{q}$ and $\hat{p}$ within the Hamiltonian $H(\hat{q},\hat{p})$. At the path integral level the different orderings correspond to different possible discretizations [Sakita (1985)]. Before analyzing what happens in the CPI, let us notice that also in quantum mechanics no ordering problem arises if we consider Hamiltonians of the form $H(q,p) = p^2/2 + V(q)$. There are instead problems when we couple q and p within the argument of the Hamiltonian H. For example the two expressions

$$\hat{p}^2\hat{q}^2 + \hat{q}^2\hat{p}^2 + \hat{q}\hat{p}^2\hat{q}, \qquad \hat{p}\hat{q}\hat{p}\hat{q} + \hat{q}\hat{p}\hat{q}\hat{p} + \hat{p}\hat{q}^2\hat{p} \qquad (I.1)$$

are two different Hermitian operators associated with the same classical observable $3q^2p^2$. Using the fundamental commutator $[\hat{q},\hat{p}] = i\hbar$ it is quite easy to prove that

$$\hat{p}^2\hat{q}^2 + \hat{q}^2\hat{p}^2 + \hat{q}\hat{p}^2\hat{q} = \hat{p}\hat{q}\hat{p}\hat{q} + \hat{q}\hat{p}\hat{q}\hat{p} + \hat{p}\hat{q}^2\hat{p} - \hbar^2,$$

which implies that, even if they are both Hermitian and associated with the same classical observable, the two operators of Eqn.(I.1) are not equivalent but differ by $\hbar^2$ terms.

In the CPI $\hat{q}$ and $\hat{p}$ commute. Nevertheless $\widehat{\varphi}^a$ and $\widehat{\lambda}_a$ do not commute and one should analyze whether there are ordering problems. Let us limit ourselves to the bosonic part of the theory. The bosonic part of the evolution operator appearing in the weight of the CPI is the Liouvillian $\widehat{L} = \lambda_a\omega^{ab}\partial_b H$. If we consider a Hamiltonian $H(q,p) = q^np^m$, then the

Path Integrals for Pedestrians

terms of the Liouvillian can be ordered in the following different ways:

$$\hat{L} = m \left[\alpha_1 \lambda_q q^n p^{m-1} + \alpha_2 q \lambda_q q^{n-1} p^{m-1} + \ldots + \alpha_{n+1} q^n \lambda_q p^{m-1} \right]$$
$$- n \left[\beta_1 \lambda_p p^m q^{n-1} + \beta_2 p \lambda_p p^{m-1} q^{n-1} + \ldots + \beta_{m+1} p^m \lambda_p q^{n-1} \right], \text{(I.2)}$$

where the sum of the weights α_j and β_j is normalized as $\sum_{j=1}^{n+1} \alpha_j = \sum_{j=1}^{m+1} \beta_j = 1$.
If we calculate the Hermitian conjugate of Eqn.(I.2) then we get:

$$\hat{L}^\dagger = m \left[\alpha_1 q^n \lambda_q p^{m-1} + \alpha_2 q^{n-1} \lambda_q q p^{m-1} + \ldots + \alpha_{n+1} \lambda_q q^n p^{m-1} \right]$$
$$- n \left[\beta_1 p^m \lambda_p q^{n-1} + \beta_2 p^{m-1} \lambda_p p q^{n-1} + \ldots + \beta_{m+1} \lambda_p p^m q^{n-1} \right] \text{(I.3)}$$

where we have used the fact that with the standard scalar product, $\langle \psi | \tau \rangle = \int d\varphi \, \psi^*(\varphi) \tau(\varphi)$, both $\hat{\varphi}$ and $\hat{\lambda}$ are Hermitian operators. Since the ordering (I.2) has to guarantee the Hermiticity of the Liouvillian $\hat{L}$, we must impose that $\hat{L}^\dagger - \hat{L} = 0$. Usinge Eqns.(I.2) and (I.3) Hermiticity is equivalent to the following equation among the coefficients α_j and β_j:

$$m \sum_{j=1}^{n+1} \alpha_j [n - 2(j-1)] = n \sum_{j=1}^{m+1} \beta_j [m - 2(j-1)]. \qquad \text{(I.4)}$$

If the coefficients α_j and β_j satisfy the previous equation then the Liouvillian (I.2) is Hermitian with the associated ordering. An example of Hermitian ordering is the one for which the coefficients satisfy: $\alpha_j = \alpha_{n-j+2}$ and $\beta_j = \beta_{m-j+2}$. This ordering generalizes the Weyl one in which all the coefficients α and β are equal: $\alpha_j = \frac{1}{n+1}$ and $\beta_j = \frac{1}{m+1}$. The pre-point ordering $\alpha_j = \delta_{j,1}$ and $\beta_j = \delta_{j,1}$ satisfies Eqn.(I.4) and corresponds to the Liouvillian in which all the operators λ are on the left of the operators φ, i.e. $\hat{L} = \lambda_a \omega^{ab} \partial_b H$. The end-point ordering corresponds instead to the following choice of the coefficients: $\alpha_j = \delta_{j,n+1}$ and $\beta_j = \delta_{j,m+1}$ which satisfies Eqn.(I.4) and produces the Liouvillian $\hat{L} = \omega^{ab} \partial_b H \lambda_a$ with all the operators λ on the right of the operators φ.

Before moving on, let us notice that, besides Eqn.(I.4), which guarantees the Hermiticity of the Liouvillian, the coefficients α_j and β_j must satisfy also the condition of normalization: $\sum_j \alpha_j = \sum_j \beta_j = 1$ which can also be written as

$$mn \sum_{j=1}^{n+1} \alpha_j = mn \sum_{j=1}^{m+1} \beta_j. \qquad \text{(I.5)}$$

Taking the difference between Eqns.(I.5) and (I.4) we get:

$$m \sum_{j=1}^{n+1} \alpha_j (j-1) = n \sum_{j=1}^{m+1} \beta_j (j-1). \tag{I.6}$$

The previous condition is crucial for proving that, even if there is more than one possible Hermitian Liouvillian, all the Hermitian Liouvillians are equivalent to the pre-point one $\widehat{L} = \lambda_a \omega^{ab} \partial_b H$. In fact let us rewrite the Liouvillian $\widehat{L}$ in a compact form as:

$$\widehat{L} = m \sum_{j=1}^{n+1} \alpha_j q^{j-1} \lambda_q q^{n-j+1} p^{m-1} - n \sum_{j=1}^{m+1} \beta_j p^{j-1} \lambda_p p^{m-j+1} q^{n-1}. \tag{I.7}$$

If we use the commutator $[\varphi^a, \lambda_b] = i\delta^a_b$ then we can rewrite Eqn.(I.7) as

$$
\begin{aligned}
\widehat{L} = {}& m \sum_{j=1}^{n+1} \alpha_j \lambda_q q^n p^{m-1} + m \sum_{j=1}^{n+1} i\alpha_j (j-1) q^{j-2+n-j+1} p^{m-1} \\
& - n \sum_{j=1}^{m+1} \beta_j \lambda_p p^m q^{n-1} - n \sum_{j=1}^{m+1} i\beta_j (j-1) p^{j-2+m-j+1} q^{n-1} \\
= {}& m\lambda_q q^n p^{m-1} - n\lambda_p p^m q^{n-1} \\
& + iq^{n-1} p^{m-1} \left(m \sum_{j=1}^{n+1} \alpha_j (j-1) - n \sum_{j=1}^{m+1} \beta_j (j-1) \right).
\end{aligned}
$$

Using Eqn.(I.6) it turns out the last term above is zero, so we can conclude that the most general Hermitian Liouvillian $\widehat{L}$ associated to the Hamiltonian $H = q^n p^m$ is equivalent to the pre-point Liouvillian in which all the operators λ are on the left of the operators φ. This proves that all the Liouvillians are equivalent and that there is no ordering problem for what concerns the bosonic part of the CPI.

The discussion regarding the Grassmann part of the CPI is more complicated, due to the fact that there is more than one possible scalar product, as explained in [Deotto *et al.* (2003a,b)]. Without entering into further details we can say that the only scalar products that assure the Hermiticity of the Grassmann part of the Hamiltonian $\widetilde{\mathcal{H}}$, independently from the particular physical system that we consider, are the gauge and the symplectic scalar products defined in [Deotto *et al.* (2003a,b)]. In these cases there are only two possible orderings, which guarantee the Hermiticity of the Hamiltonian:

$$\widetilde{\mathcal{H}}_{G,1} = i\bar{c}_a \omega^{ab} \partial_b \partial_d H c^d, \qquad \widetilde{\mathcal{H}}_{G,2} = -ic^d \omega^{ab} \partial_b \partial_d H \bar{c}_a.$$

They are completely equivalent, as we can easily prove using the graded commutators of the theory:

$$-ic^d\omega^{ab}\partial_b\partial_d H\bar{c}_a = i\bar{c}_a\omega^{ab}\partial_b\partial_d H c^d - i\omega^{ab}\partial_b\partial_d H\left[\bar{c}_a, c^d\right] =$$
$$= i\bar{c}_a\omega^{ab}\partial_b\partial_d H c^d - i\omega^{db}\partial_b\partial_d H = i\bar{c}_a\omega^{ab}\partial_b\partial_d H c^d.$$

So we can conclude that for what concerns the Grassmann part of the Hamiltonian $\widetilde{\mathcal{H}}$ there is more than one possible Hermitian ordering but also in this case the associated Hamiltonians are equivalent and no ordering problem arises.

Bibliography

Abraham, R. and Marsden, J. E. (1978). *Foundations of Mechanics* (Benjamin Cummings, New York).

Abrikosov Jr., A. A. and Gozzi, E. (2000). *Nucl. Phys. Proc. Suppl.* **88**, p. 369, eprint quant-ph/9912050.

Abrikosov Jr., A. A., Gozzi, E., and Mauro, D. (2003). *Mod. Phys. Lett A* **18**, p. 2347, eprint quant-ph/0308101.

Abrikosov Jr., A. A., Gozzi, E., and Mauro, D. (2005). *Annals Phys.* **317**, p. 24, eprint quant-ph/0406028.

Albeverio, S. and Hoegh-Krohn, R. (1976). *Mathematical Theory of Feynman Path Integrals*, Vol. 523 (Lectures Notes in Mathematics, Springer-Verlag, Berlin and Heidelberg and New York).

Baker, G. A. (1958). *Phys. Rev. D* **109**, p. 2198.

Bayen, F., Flato, M., Fronsdal, C., Lichnerowicz, A., and Sternheimer, D. (1978a). *Annals Phys.* **111**, p. 61.

Bayen, F., Flato, M., Fronsdal, C., Lichnerowicz, A., and Sternheimer, D. (1978b). *Annals Phys.* **111**, p. 111.

Belavin, A., Poyakov, A., Schwartz, A., and Tuytin, Y. (1975). *Phys. Lett. B* **59**, p. 85.

Berezin, F. A. (1981). *Sov. Phys. Usp.* **23**, p. 763.

Berezin, F. A. and Marinov, M. S. (1977). *Annals Phys.* **104**, p. 336.

Bopp, F. (1961). *Werner Heisenberg und die Physik unserer Zeit* (Springer Verlag, Berlin).

Carta, P., Gozzi, E., and Mauro, D. (2006). *Annalen Phys.* **15**, p. 177, eprint hep-th/0508244.

Casalbuoni, R. (1976). *Nuovo Cim. A* **33**, p. 389.

Cattaruzza, E. and Gozzi, E. (2012). *Phys. Lett. A* **376**, p. 3017, eprint 1207.5706.

Cattaruzza, E., Gozzi, E., and Neto, A. F. (2011). *Annals Phys.* **326**, p. 2377, eprint 1010.0818.

Coleman, S. (1985). *Aspects of Symmetry: Selected Erice Lectures of Sidney Coleman* (Cambridge University Press, New York).

Das, A. (1993). *Field Theory: a Path Integral Approach* (World Scientific, Singapore).

Deotto, E. and Gozzi, E. (2001). *Int. J. Mod. Phys. A* **16**, p. 2709, eprint hep-th/0012177.

Deotto, E., Gozzi, E., and Mauro, D. (2003a). *J. Math. Phys.* **44**, p. 5902, eprint quant-ph/0208046.

Deotto, E., Gozzi, E., and Mauro, D. (2003b). *J. Math. Phys.* **44**, p. 5937, eprint quant-ph/0208047.

Dewitt, B. S. (1992). *Supermanifolds*, Cambridge monographs on mathematical physics (Cambridge University Press, Cambridge, New York).

Dirac, P. A. M. (1933). *Phys. Zeit. der Sowietunion* **5**, p. 1.

Feynman, R. P. (1948). Space-time approach to non-relativistic quantum mechanics, *Rev. Mod. Phys.* **20**, 367.

Feynman, R. P. (2005). *Feynman's Thesis: A New Approach to Quantum Theory* (World Scientific, Singapore).

Feynman, R. P. and Hibbs, A. R. (1965). *Quantum Mechanics and Path Integrals* (McGraw-Hill, New York).

Fonda, L. and Ghirardi, G. (1970). *Symmetry Principles in Quantum Physics* (Marcel Dekker, Inc., New York).

Goldstein, H. (2002). *Classical Mechanics*, 3rd edition (Addison Wesley, San Francisco).

Gozzi, E. (1984). *Phys. Rev. D* **30**, p. 1218.

Gozzi, E. (1985). *Phys. Rev. D* **31**, p. 441.

Gozzi, E. (1993). *Prog. Theor. Phys. Suppl.* **111**, p. 115.

Gozzi, E. (1994). *Chaos Solitons Fractals* **4**, p. 653.

Gozzi, E. (1995). Unpublished notes.

Gozzi, E. and Mauro, D. (2000). *J .Math .Phys.* **41**, p. 1916, eprint hep-th/9907065.

Gozzi, E. and Mauro, D. (2002). *Annals Phys.* **296**, p. 152, eprint quant-ph/0105113.

Gozzi, E., Mauro, D., and Silvestri, A. (2005). *Int. J. Mod. Phys. A* **20**, p. 5009, eprint hep-th/0410129.

Gozzi, E. and Pagani, C. (2010). *Phys. Rev. Lett.* **105**, 15, p. 150604.

Gozzi, E. and Penco, R. (2011). *Annals Phys.* **326**, p. 876, eprint 1008.5135.

Gozzi, E. and Regini, M. (2000). *Phys. Rev. D* **62**, p. 067702.

Gozzi, E. and Reuter, M. (1989). *Phys.Lett. B* **233**, p. 383.

Gozzi, E. and Reuter, M. (1993). *Phys. Rev. E* **47**, p. 726.

Gozzi, E. and Reuter, M. (1994a). *Int. J. Mod. Phys. Lett. A* **9**, 13, p. 2191.

Gozzi, E. and Reuter, M. (1994b). *Chaos Solitons Fractals* **4**, p. 1117.

Gozzi, E. and Reuter, M. (1995). Unpublished notes .

Gozzi, E., Reuter, M., and Thacker, W. (1992a). *Chaos Solitons Fractals* **2**, p. 441.

Gozzi, E., Reuter, M., and Thacker, W. D. (1989). *Phys. Rev. D* **40**, p. 3363.

Gozzi, E., Reuter, M., and Thacker, W. D. (1992b). *Phys. Rev. D* **46**, p. 757.

Khandekar, D. C., Lawande, S. V., and Bhagwat, K. V. (1993). *Path Integral Methods and Their Application* (World Scientific, Singapore).

Kleinert, H. (1990). *Path Integrals in Quantum Mechanics, Statistics and Polymer Physics* (World Scientific, Singapore).

Koopman, B. O. (1931). *Proc. Natl. Acad. Sci. USA* **17**, p. 315.

Kubo, R. (1964). *Jour. Phys. Soc. Jap.* **19**, p. 2127.

Littlejohn, R. G. (1986). *Phys. Rep.* **138**, 193.

Marinov, M. S. (1991). *Phys. Lett. A* **153**, p. 5.

Moyal, J. E. (1949). *Proc. Cambr. Phil. Soc.* **45**, p. 99.

Rivers, R. J. (1987). *Path Integral Methods in Quantum Field Theory* (Cambridge University Press, New York).

Roman, P. (1965). *Advanced Quantum Theory* (Addison Wesley, San Francisco).

Sakita, B. (1985). *Quantum Theory of Many Variables and Fields* (World Scientific, Singapore).

Schulman, L. S. (1981). *Techniques and Applications of Path Integration* (Wiley Publ., New York, NY).

Sharan, P. (1979). *Phys. Rev. D* **20**, p. 414.

Surin, T.-N. and Kurchan, J. (2004). *Jour. Stat. Phys.* **116**, p. 1201.

Swanson, M. S. (1992). *Path Integrals and Quantum Processes* (Academic Press, Boston).

Thacker, W. (1997a). *Jour. Math. Phys.* **38**, p. 2389.

Thacker, W. (1997b). *Jour. Math. Phys.* **38**, p. 300.

von Neumann, J. (1932a). *Ann. Math.* **33**, p. 587.

von Neumann, J. (1932b). *Ann. Math.* **33**, p. 789.

West, P. C. (1986). *Introduction to Supersymmetry and Supergravity* (World Scientific, Singapore).

Weyl, H. (1927). *Z. Phys.* **46**, p. 1.

Wick, G., Wightman, A., and Wigner, E. (1952). *Phys. Rev.* **88**, p. 101.

Wiegel, F. W. (1986). *Introduction to Path Integrals Methods in Physics and Polymer Science* (World Scientific, Singapore).

Wigner, E. P. (1932). *Phys. Rev.* **40**, p. 749.

Woodhouse, N. (1997). *Geometric Quantization* (Clarendon Press, Oxford).

Zinn-Justin, J. (1996). *Quantum Field Theory and Critical Phenomena* (University Press, Oxford).

Zinn-Justin, J. (2005). *Path Integral in Quantum Mechanics* (Oxford University Press, Oxford).

Printed in the United States
By Bookmasters